A Practical Guide to Renewable Energy

Microgeneration Systems and their Installation

A Practical Guide to Renewable Energy Microgeneration Systems and their Installation

Christopher Kitcher

Routledge
Taylor & Francis Group
LONDON AND NEW YORK

First edition published 2012
by Routledge
2 Park Square, Milton Park, Abingdon, Oxon OX14 4RN

Simultaneously published in the USA and Canada
by Routledge
711 Third Avenue, New York, NY 10017

Routledge is an imprint of the Taylor & Francis Group, an informa business

British Library Cataloguing in Publication Data
A catalogue record for this book is available from the British Library

Library of Congress Cataloguing in Publication Data

Kitcher, Chris.

A practical guide to renewable energy : power systems and their installation / Chris Kitcher. — 1st ed.

Summary: "Learn more about renewable energy, how to install and inspect renewable energy systems and gain certification. This is a perfect introduction to one of the construction industry's leading growth areas. It provides an overview of all types of renewable energy sources, as well as information relating to the installation and inspection of renewable energy systems. The practical focus in this book will give you the confidence to pass micro-generation exams, discuss the subject with clients and work on all new and emerging renewable energy systems. It does this by providing you with: Step-by-step instructions in how to fit and test renewable energy systems Clear diagrams, photos and flow charts that demonstrate core principles Questions and answers that enable you to test your knowledge and further your understanding of the subject As a student or professional this textbook will provide the information needed to pass your course and is also an ideal onsite reference. Chris Kitcher is an Electrical Installation lecturer at Central Sussex College, author of the bestselling Practical Guide to Inspection, Testing and Certification of Electrical Installations and has 49 years of experience in the electrical industry"— Provided by publisher.

1. Electric power systems. 2. Renewable energy sources. I. Title.
TK1001.K543 2012
621.042—dc23
2011039400

ISBN: 978-0-08-097064-6 (pbk)
ISBN: 978-0-08-097065-3 (ebk)

Typeset in Helvetica
by RefineCatch Limited, Bungay, Suffolk
Printed and bound in Great Britain by Bell & Bain Ltd., Glasgow

Contents

Acknowledgements

The author would like to thank the following people and companies who provided information and/or permission to reproduce photographs and/or diagrams:

Sanyo
Trojan Battery Company
Shutterstock
Sovello
Nexgen Wind Monitoring Solutions
Wikipedia
MK Electric
Stelrad Radiators
Caron Alternative Energy
Widos Technology
Barnes Plastic Welding Equipment
Special Plasters
colcrete-eurodrill.com
BDR Thermea
Simon Wood, Megger UK
Central Sussex College
Cornelia Raffeiner, Renisin
IHS Energy

Preface

The reduction of greenhouse gas emissions is an agreement which has been reached worldwide. An energy package has been endorsed by the European Council with objectives that greenhouse gases are to be reduced by a minimum of 20% by 2020, possibly rising to 30% if agreement with governments from other parts of the world can be reached.

Much of the reduction in greenhouse gases can be achieved by the use of renewable energy using the power generated by the sun. Technologies have evolved over many years which allow us to turn the energy provided by the sun into electricity and heat, the energy can also be stored and used as required. As the energy from the sun is free it can be a very useful source of fuel when linked to the correct technology, unfortunately most of this technology is quite expensive and in many instances its installation is not a viable proposition as the money invested in the installation would not be recovered before the system would need replacing. However, in the UK the government has introduced a system whereby the owner of the installation is paid a tariff for each unit of energy produced by the system. This of course will reduce the recovery time of the cost of the installation quite considerably.

Although these technologies have been around for many years the use of them in domestic installations has been very rare and up until recently the technology had not advanced very much. The incentives offered by the government along with the rising cost of fossil fuels has increased the investment in new technology and this, along with the commitment to reduce greenhouse gases, means that more and more dwellings will be fitted with energy saving technologies.

Schemes have been set up to ensure that these technologies are installed to a standard which is acceptable, and in most cases if not all, any feed in tariffs will only be available for installations which have been installed by an installer who is registered with an organisation which provides a Microgeneration Certification Scheme (MCS). Any technologies installed by a member of one of these organisations will be of a good standard with all materials being MCS recognised. Figures 1 and 2 are the logos which registered installers along with suppliers of equipment would use to show that they are MCS registered.

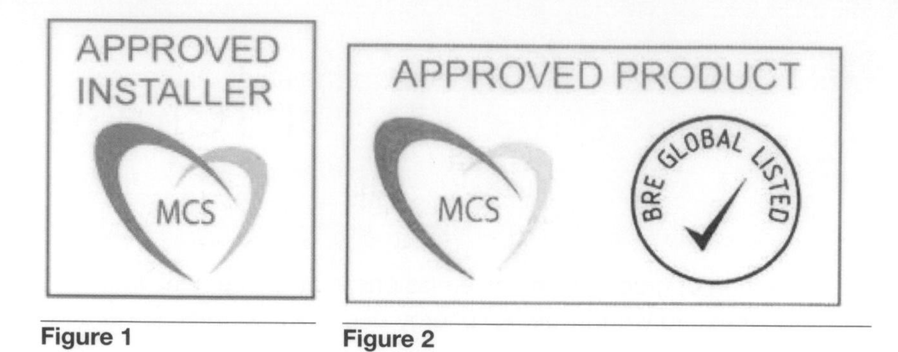

Figure 1 **Figure 2**

This book is intended as an introduction and practical guide to the renewable technologies which are becoming more and more popular. It describes what the technologies are, how they work, the different methods of installation and the types of materials used.

Chris Kitcher

CHAPTER 1

Solar photovoltaic systems

The combination of the Greek word for light, which is photos and the term used for a unit of electromotive force, which is volt is combined to form the word photovoltaic. This is used as the name for a system which converts the energy, provided by the sun (light) directly into electricity by using a solid state device which is called a photovoltaic cell.

The sun supplies us with energy in the form of solar radiation. The intensity of this solar radiation is known as irradiance and is measured in watts per metre square (W/m^2). In good weather when the sun is at its peak the irradiance may reach 1000Wm^2. Irradiance is also referred to as insolation (not to be confused with insulation) which is irradiance × time and the amount of irradiance which reaches the earth is made up of direct insolation, and diffuse insolation.

Direct insolation is the amount of solar irradiance measured at a given point on earth perpendicular to the sun's rays, and diffuse insolation is the amount of radiation which reaches the same point but which has been reflected or scattered by elements which make up the earth's atmosphere. It is thought that around 60% of the solar radiation which reaches the UK is diffuse radiation although this alters depending on the amount of cloud. Clearly on a fine bright day with little or no cloud most of the radiation would be direct, whereas in the winter and on cloudy days the radiation would be almost entirely diffuse. Over a year this averages out at 60% diffuse and 40% direct.

It is the job of the photovoltaic cell to capture and convert into electrical energy the solar radiation which falls on to it.

These cells are generally constructed using silicone although other materials are being developed. Silicon is the most abundant element on earth after oxygen, and of course this makes it available in almost unlimited quantities. For use in the electronics industry silicon has to be very pure and the process of purifying the silicone is very expensive. The silicon used in the solar industry does not have to be as pure as that used to make semiconductors and for this reason much of the silicon used is the waste from the semiconductor industry, this of course helps to reduce the cost. Unfortunately with the increased demand for photovoltaic panels there is not enough waste to fulfil the demand and the shortfall has to be made up of pure material.

As with anything else, due to increased demand new methods of production will be developed which in turn will reduce the cost of the materials.

There are four common types of silicon photovoltaic cells in use and these are:

- Mono-crystalline (single crystal) silicon cells
- Polycrystalline silicone cells
- Ribbon pulled silicon
- Amorphous silicon.

Mono-crystalline cells (c-Si)

Mono-crystalline cells are the most expensive as they are formed of pure silicon, they are also the most efficient with an efficiency of 15 to 18%.

These cells are sliced thinly from round rods of silicon (Fig. 1.1a/b), but if they are fitted into panels in this shape there will be a large area of the panel which will not be producing energy (Fig. 1.2); for this reason they are rarely used. It is possible to form the unit holding the cells out of transparent materials and in some instances where light is required.

In an area below the module the use of round cells would be an option.

It is better for energy production if the cells are cut to form a square but of course this would produce a lot of unusable waste (Fig. 1.3).

Figure 1.1 (a) Silicone after it is cut. (b) Silicone before it is cut.

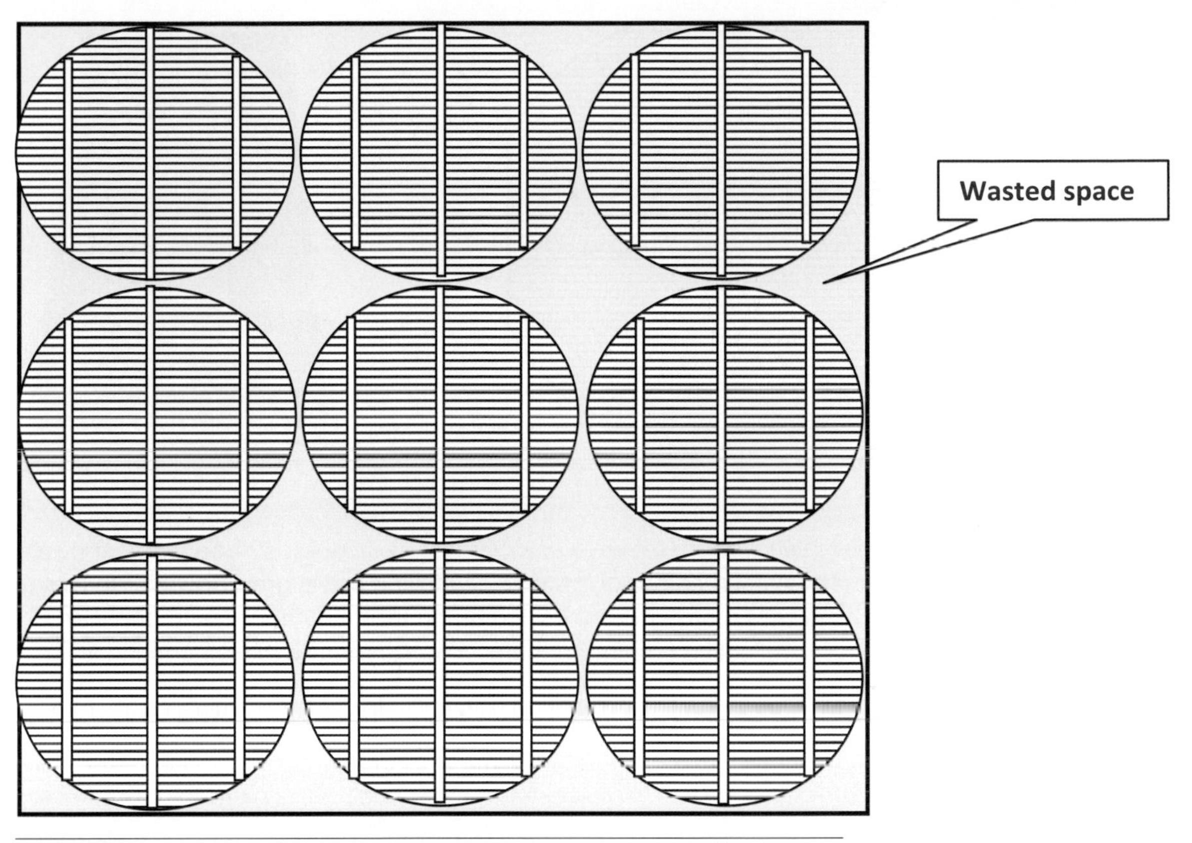

Figure 1.2 A number of round cells fitted side-by-side onto a panel with a large area of panel not producing energy.

Using square cells would certainly allow the modules to produce more energy but the cost of the cells would rise due to the amount of silicon which would need to be discarded.

A better option, and one that is often used with mono-crystalline cells, is to partially trim the round cells (Fig. 1.4a/b), this of course reduces the amount of waste silicon and unused area on the solar module which in turn helps keep the manufacturing costs down while making the best use of the available area.

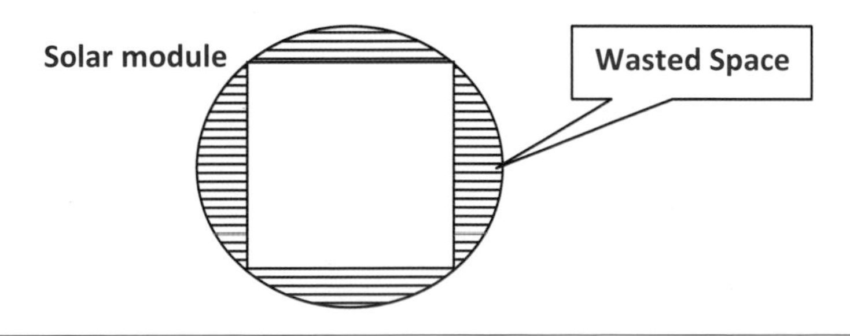

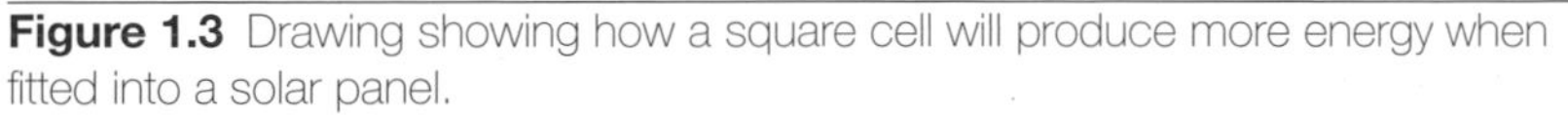

Figure 1.3 Drawing showing how a square cell will produce more energy when fitted into a solar panel.

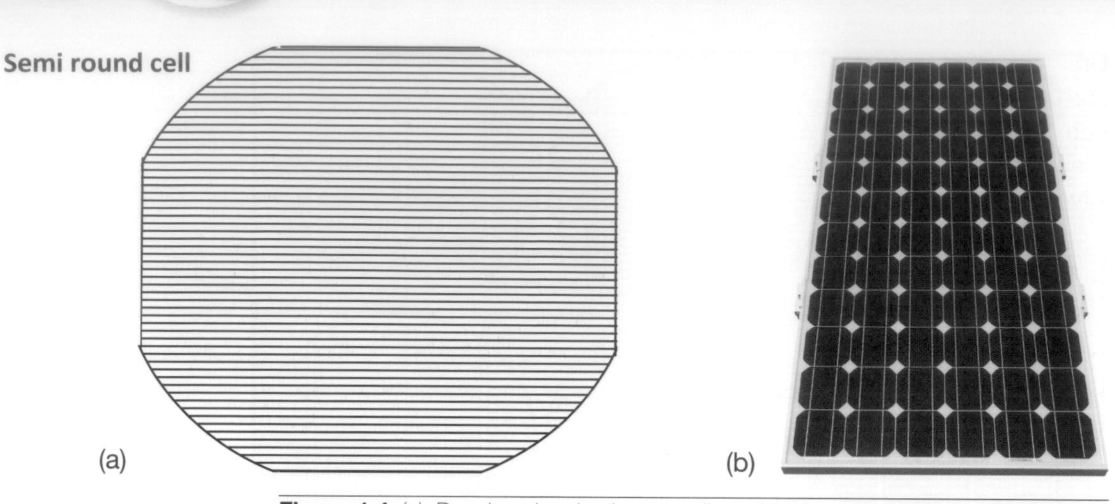

Figure 1.4 (a) Drawing showing how a cell can be trimmed. (b) Trimmed cells fitted into a panel.

The natural colour of a mono-crystalline cell is grey, however to make the cells more efficient it is common for manufacturers to apply an anti-reflective coating to them. When an anti-reflective coating is applied the colour is uniform and will be from black to very dark blue. Each cell will usually be between 100mm^2 and 150mm^2 with a thickness of 0.2 to 0.3mm.

Polycrystalline silicon cells (c-Si)

Polycrystalline cells are made by melting the material and casting it into large blocks. When the blocks cool down they form a solid block of silicon crystals. These blocks are then cut into square bars (Fig. 1.5) from which thin slices of silicon are cut, and these are usually around 3mm thick and can be from 100mm^2 to 200mm^2. Using larger cells brings down the costs as each module will need fewer cells. The efficiency of this type of cell is from 13 to 16%.

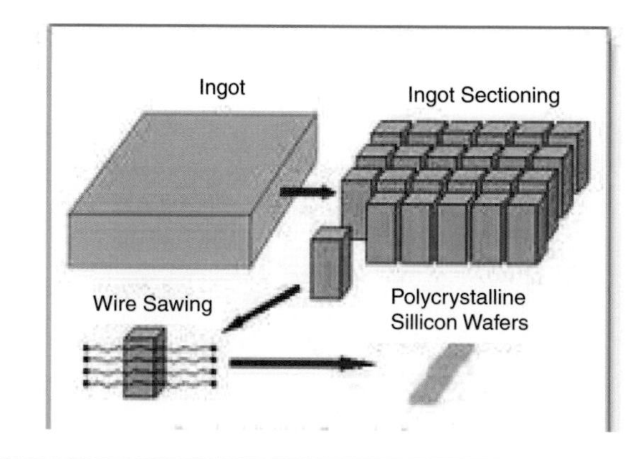

Figure 1.5 Process/sequence for cutting polycrystalline cells.

Unlike the mono-crystalline cells which have a smooth uniform surface, polycrystalline cells reflect light differently and the crystals can be clearly seen. The natural colour of these cells is blue unless they have an anti-reflective coating, in which case they will be a silver-grey colour.

Ribbon pulled silicon cell

Conventional methods used to manufacture the silicon wafers used in the production of photovoltaic cells has high material losses often up to 40%. Due to this a lot of technological development has been carried out and various ribbon pulling processes have been developed.

This process involves the silicon film being pulled directly out of the molten silicon melt (Fig. 1.6) at the correct thickness for use as a silicon wafer. After the ribbons have been formed they are cut to the required size using a laser. Using this method the material wastage is very small, possibly less than 10%.

The efficiency of this type of cell is around 14%. Other technologies have been developed to produce thin wafers of silicon which in time will/should reduce the cost of PV modules.

Amorphous silicon (a-Si)

Amorphous silicon cells are much cheaper to produce than those made from crystalline silicon, this type of cell also absorbs light better and for that reason can be made thinner (Fig. 1.7).

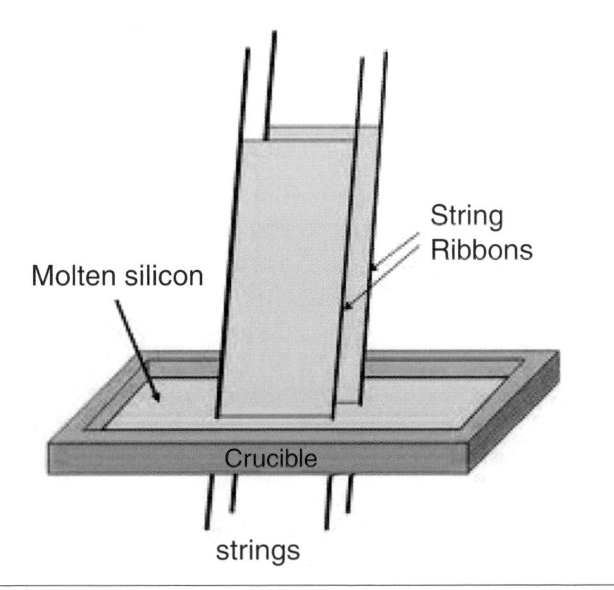

Figure 1.6 The process for the manufacture of a ribbon pulled silicon cell.

Figure 1.7 Amorphous silicon cell.
Source: Sanyo

The manufacturing process operates at a much lower temperature than that required to produce the other types of cell, and this in turn helps to reduce the cost. At the present time these cells only have an efficiency of around 6% and they are widely used for solar powered toys and calculators. Efficiency is not an issue with these types of products but low cost is very important.

To ensure that the cells operate at the best possible efficiency an anti-reflective coating is applied (this would be either titanium dioxide or silicon nitride). It is this coating which changes the colour of the cell from dark grey (mono-crystalline) or silver-grey (polycrystalline) to black through to dark blue for mono-crystalline cells to a lighter blue for polycrystalline cells.

For decorative purposes other colours can be produced, these are gold, violet, green and brown. The cells can also be used without the reflective coating if required to blend in with other building materials, but of course they will be less efficient due to the possibility of around 30% of sunlight being reflected off of the surface and they will therefore be less usable.

PV modules

Due to the size of a solar cell the output of a single cell is very small. Because of this the individual cells have to be connected in series and parallel to form a PV module which will produce a suitable output whilst remaining at a physical size and weight for ease of installation.

The actual size of the panel depends on how many solar cells are used in its manufacture. It is usual to have 36 or 72 cells (Fig. 1.8) although this can change depending on the size of the cells, the manufacturer and the output required from each module.

A collection of PV cells in one unit is referred to as a PV module and a collection of modules is referred to as a PV array.

The rated output of a PV module is given in Wp, this is the maximum peak output in watts which an individual module could produce. For a total installation the rated output is given as kWp, this is the total maximum peak output in kilowatts which an installation could produce.

As described earlier, the maximum amount of energy which falls on to an area 1m^2 is calculated to be 1000w or 1kW. This value is used by manufacturers to calculate the peak power output of their modules and all manufacturers must produce data showing the output of their PV modules which have been tested using what are called standard

Figure 1.8 A typical solar panel installation.

test conditions (stc). These are 1000W/m^2, a temperature of 25°C and an air mass of 1.5. This is to ensure that the user/purchaser can compare different products fairly.

Of course the output given by a manufacturer is rarely reached as it is a maximum value and the amount of solar radiation can change very quickly. Although each module has a peak output given by the manufacturer, it could only be attained on a very fine day and even then only for a very short period of time. It should be remembered that a PV cell cannot store any energy and for that reason the output of a cell can and does change very quickly. A small cloud passing in between the sun and the PV cell could reduce the output of a cell to virtually nothing almost instantly.

The amount of time the sun is visible is also a major factor in the output of PV installations and we are all aware that in the UK we get far more sun in the summer months than we do in the winter. Below is a table which shows the average kWh/m^2 each month in the South-East of England

Jan	Feb	Mar	Apr	May	Jun
0.78	1.52	2.3	4.23	5.34	5.64

Jul	Aug	Sep	Oct	Nov	Dec
5.55	4.79	3.32	1.95	1.08	0.68

In round figures a 1m^2 solar panel will produce a little more than 100w of energy, therefore to have a solar array which would produce an output of 1kWp we would need 10 panels which in reality would provide between 600 and 900kWh per year. (These figures will undoubtedly improve as new technology becomes available.) As the table above demonstrates, the average energy level in January is less than one fifth of the energy level in June or July.

The siting of the PV array is also very important. Although they convert light to energy and will work when exposed to any light, direct sunlight gives far better results than reflected light. For that reason the panels should be pointed directly at the sun so that as much sunlight as possible is collected. As a practical exercise this is virtually impossible in most cases as the sun's angle and position changes constantly throughout the year (Fig. 1.9). The greater the angle of the sun to the panel, the less intense the light becomes.

As we know the sun rises in the east and sets in the west. When talking about solar energy systems, south is generally given as 0°, east is shown as –90° and west is shown as 90°. The angle at which the sun is to south is known as the azimuth angle.

As the sun rises in the east and sets in the west the best position for a fixed panel is for it to face south, this will ensure that the sun's path

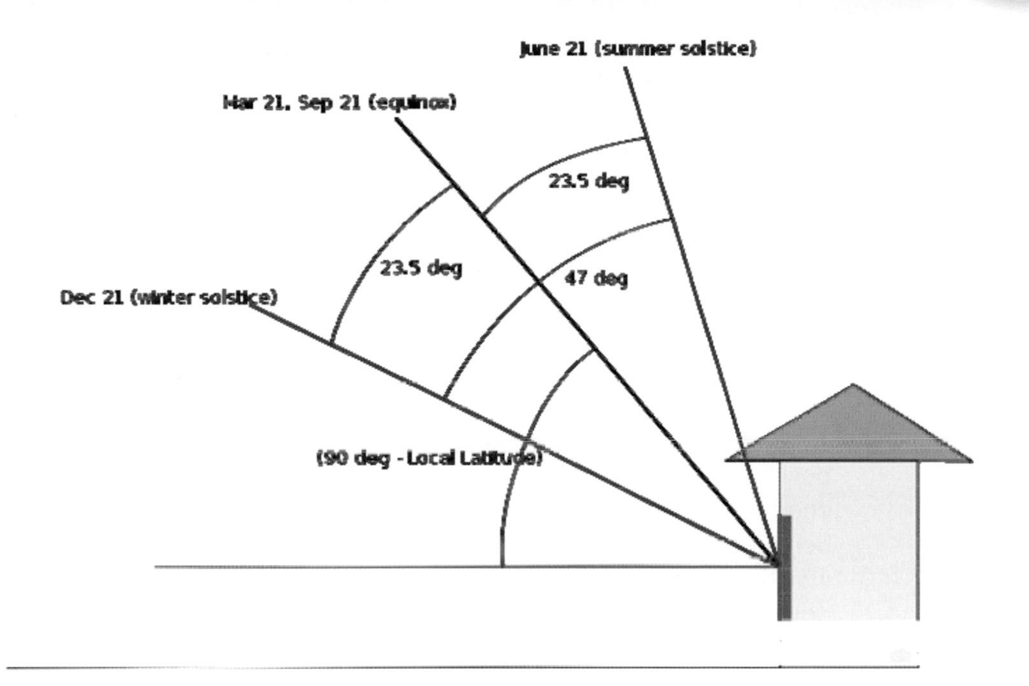

Figure 1.9 This drawing shows how the sun is at a much steeper angle during the summer months than in the winter. A panel set at around 35 degrees would be most suitable.

will move across the panel allowing it to collect energy for the entire time that the sun is in view. Sun path charts are available from the internet and all that is needed is the latitude and longitude values of the site to be entered and a chart can then be printed showing the azimuth and elevation of the sun throughout the year (Fig. 1.10).

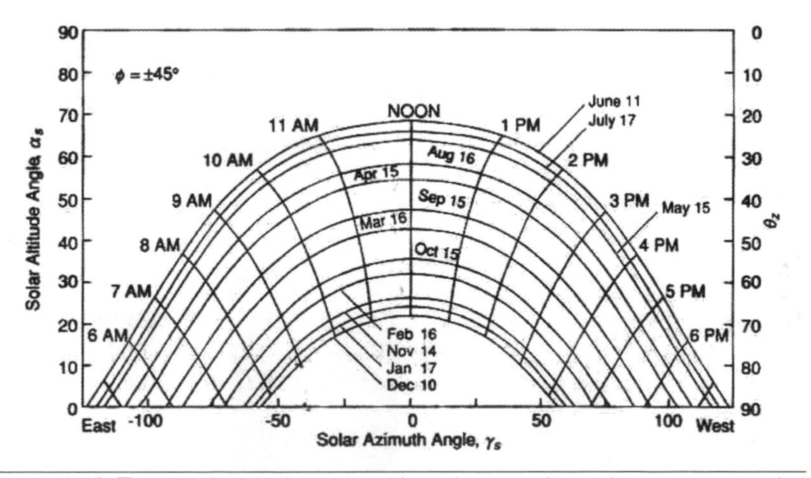

Figure 1.10 The sun rises in the east and continues to rise as it moves across the sky; it is at its highest point when it is due south at noon. As it moves west it begins to fall. Its highest point is in the summer.

Apart from moving across the sky the sun is also at a different angle to the horizontal each day of the year; it is very high in the summer and low in the winter. Obviously this means that the sun's rays will be shining on the panel from a slightly different angle each day which in turn means that the maximum output of the panel will only be obtainable on a very few days of the year.

The output can be increased by fitting a device which tracks the sun's path. It can be set to a predicted path for each day of the year and this of course will ensure that the PV panels will be facing the sun at the best angle at all times of the day. Another option is to fit a device which simply directs the PV panel at the brightest part of the sky (Fig. 1.11), this of course is normally the sun. In the United Kingdom these devices are rarely used because as we tend to get a lot of cloud the energy used to drive the tracking device would probably outweigh the benefit gained by following the sun. Another problem is that a tracker which is set to follow the brightest object in the sky would often be pointing at a bright gap between the clouds which although it may be the brightest part of the sky, would not necessarily be the sun.

Imagine the confusion on a cloudy day when a cloud passes in front of the sun but there is also lots of bright blue sky, the tracker would be moving all over the place searching out the brightest spot.

Before installing a solar PV system it is important to survey the site to get an idea of how much energy would possibly be available. There are various tools available to assist in this, and there is also an abundance of information on the internet. *www.energysavingtrust.org.uk* provides an abundance of information about all renewable technologies.

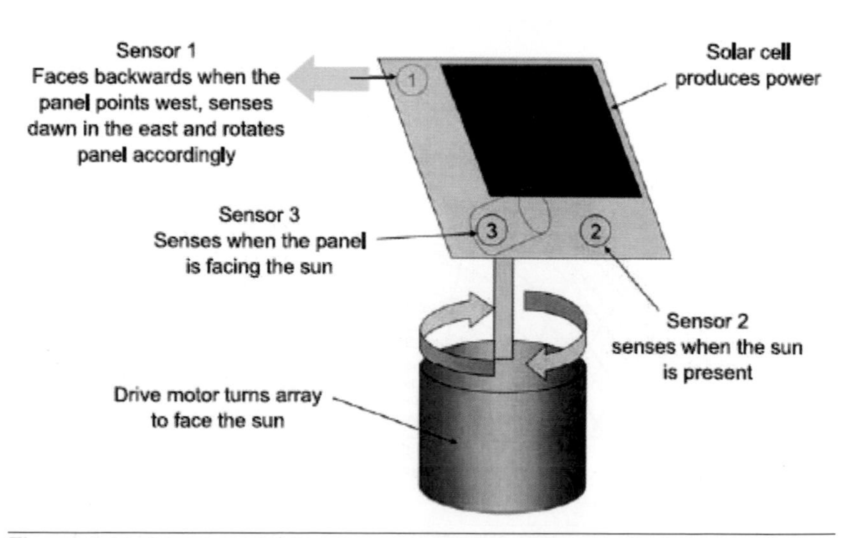

Figure 1.11 A tracker will follow the sun across the sky and will also set itself at the correct angle from the horizontal to collect as much sunlight as possible.

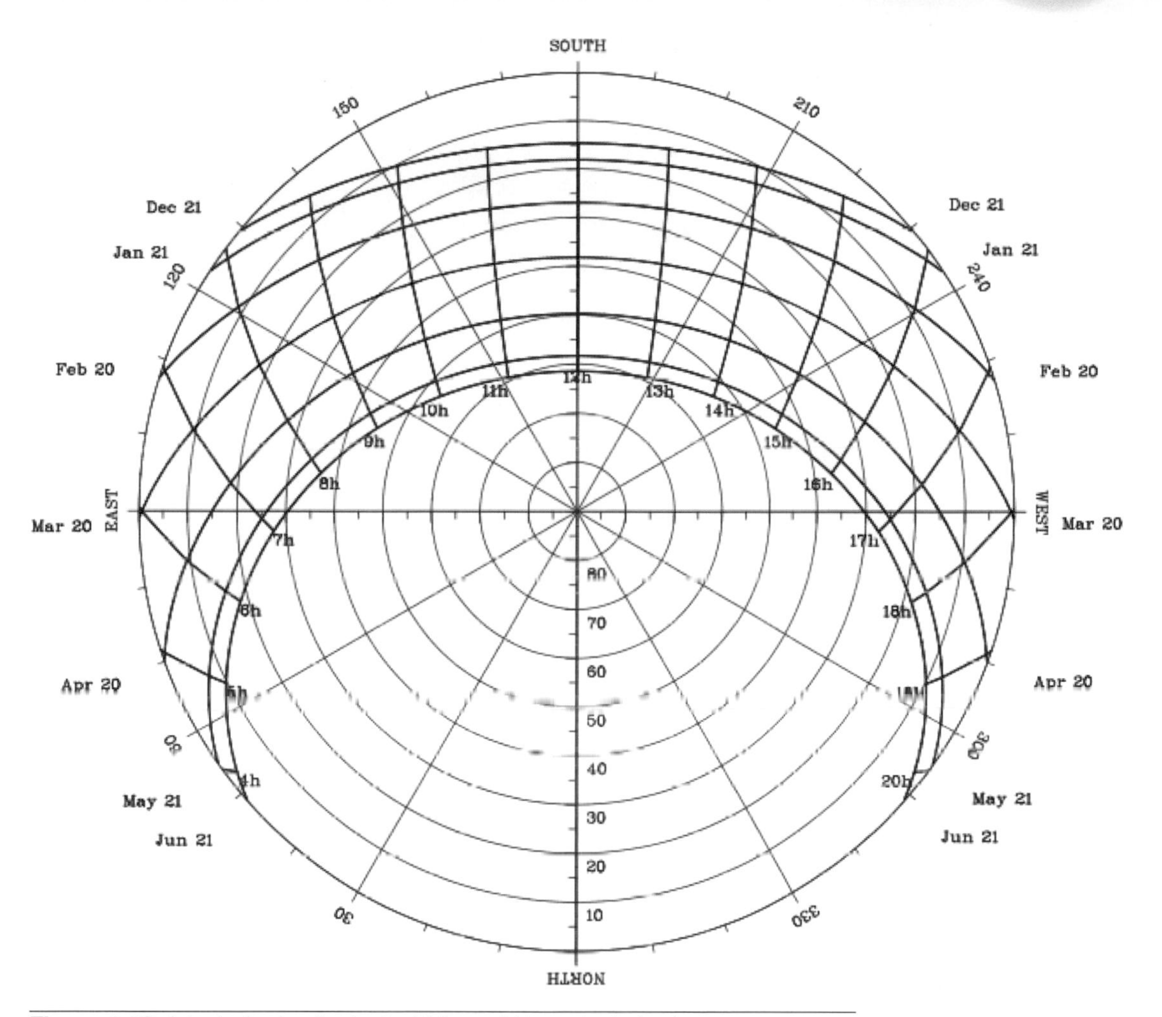

Figure 1.12 A typical azimuth chart which would be selected using map coordinates. This will show where the sun will be at any given time of the day, month and year.

The most basic equipment needed would be a compass to enable you to find out the direction which your panel would be facing, and an azimuth angle chart (Fig. 1.12).

If you set the azimuth chart so that it is facing due south it will become clear how long the sun could shine onto the panel each day.

As you would expect, a tool has been designed which will help in carrying out a site survey: it is call a solar path viewer (Fig. 1.13). The idea is that you print off a solar path chart on a piece of acetate, fix it to the viewer and then look through the chart using the eyepiece of the viewer. Not only will this allow you to calculate how long the sun's rays will be shining on to the modules, it will also indicate any objects which will block the sun's rays at different times of the day and year. This is very important, because if the sun is going to be behind a tree or a block of flats the efficiency of the system will be severely reduced.

Figure 1.13 A sun path viewer fitted with an azimuth chart to suit the building being surveyed.

Connection of PV modules

A photovoltaic module is a collection of cells linked together to form a single module, and as we have seen a module will usually consist of 72 or 36 cells. A single module will not produce a great deal of energy and for this reason modules are joined together, usually in series, to form what is called a PV string (Fig. 1.14).

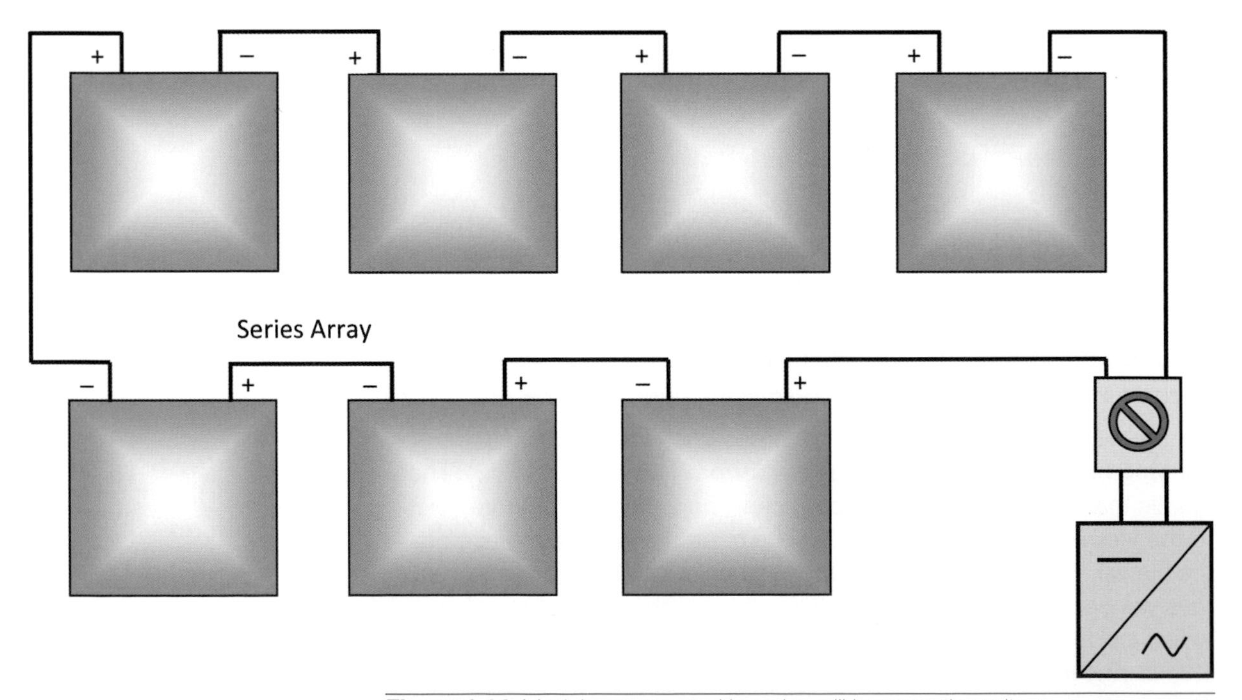

Figure 1.14 Modules connected in series will increase the voltage.

As an example, if we assume that a single module has a maximum power voltage of 17.4 volts (Vmp) and a maximum power current of 7.4A (Imp) the module would provide around 128 watts of power. This is of course is its maximum power output and would only be achieved on a very good day, also we really need a higher voltage to help us increase the efficiency when we convert our d.c. current into a.c. current for integration with the supply system.

It is far better for us if we connect a number of modules in series to increase the voltage. The connection of PV modules to increase voltage is very similar to connecting batteries to increase voltage. PV modules are polarised, in other words they have a positive and negative terminal. To connect the modules in series all that is required is that we connect the positive of one module to the negative of another until we have enough modules in a string to provide us with the voltage which we require (Fig. 1.14).

Let's say that we need a voltage of around 100 volts. To achieve this we need to calculate how many modules we should connect in series in our string.

Calculation

$$\frac{\textit{Required voltage}}{\textit{Module voltage (Vmp)}} = \textit{number of panels}$$

$$\frac{100}{17.4} = 5.74 \textit{ panels}$$

The required number of modules required in a string would be 6 which would produce a Vmp of 6 × 17.4 = 104.4 volts. Of course the current (Imp) would remain at 7.4A and to increase the current we would need to connect a number of strings in parallel (Fig. 1.15).

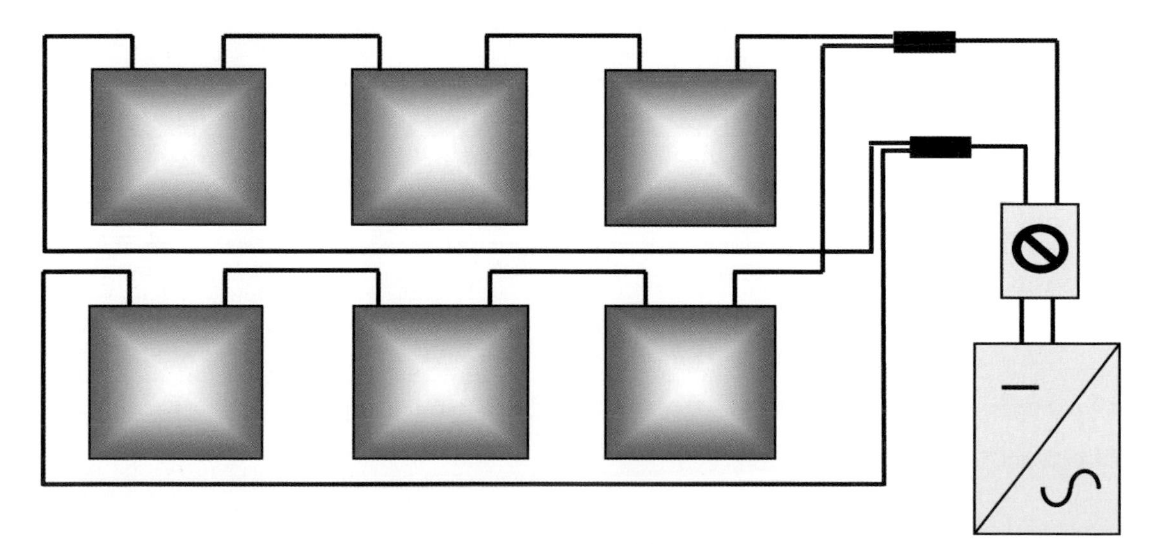

Figure 1.15 Modules connected in parallel will increase the current.

This is no different than connecting a number of batteries together in parallel to increase the current. If we required a current of around 20A we would need 3 strings and would need to connect the positive end of each string together and the negative end of each string together to form what is known as a PV array. In our example the current (Imp) would be slightly greater than 20A as 3 × 7.4 = 22.2A. This system consisting of 18 modules would now have a potential maximum power output (Pmax) of 2300 watts or 2.3Kw.

There is no limit to the number of modules and strings which can be joined to obtain a required output, but the system is usually limited by the space available which is suitable for mounting the panels onto.

The output of the PV panel will depend on its size and type, and all of the required information will be printed on the panel. Typical information is shown on the following table.

Rated maximum power	180W
Tolerance	–0/+3%
Voltage at Pmax (Vmp)	36.4V
Current at Imp (Imp)	4.95A
Open-circuit voltage (Voc)	44.2V
Short-circuit current (I sc)	5.13A
Nominal operating temp	45°C
Maximum system voltage	1000V d.c.
Maximum series fuse rating	15A
Operating temperature	–40°C to +85°C
Protection class	Module Application Class A
Cell technology type	Mono – SI
Weight (kg)	15.5
Dimensions (mm)	1580 × 808 × 45
Standard test conditions: AM1.5 E=1000W/m^2	Tc = 25°C

Types of system

A decision which has to be made at a very early in the design stage is the type of system which is going to be installed. There are three types of system which are commonly used.

Stand alone system

A stand alone system is a PV installation which either uses the energy as it is collected or stores the energy for use as required. The type of system which uses energy as it is produced is really not very practical as clearly it is only useable in daylight and the amount of energy would vary almost by the second. This system is greatly improved by the addition of a battery storage system which would allow the energy to be stored and then used on demand.

In a very basic installation the energy would be stored in batteries, usually 12 volt although this voltage could be increased or reduced depending on the requirements of the user. The energy could then be used at the stored voltage to supply equipment rated at the same voltage, this would be d.c. This type of system is fine for installations which have dedicated circuits specifically for the PV installation (Fig. 1.16a/b).

The installation would consist of the PV panels, a charge controller and the batteries used for storage. For this system to be effective it would be very important for the designer to assess the requirements of the installation.

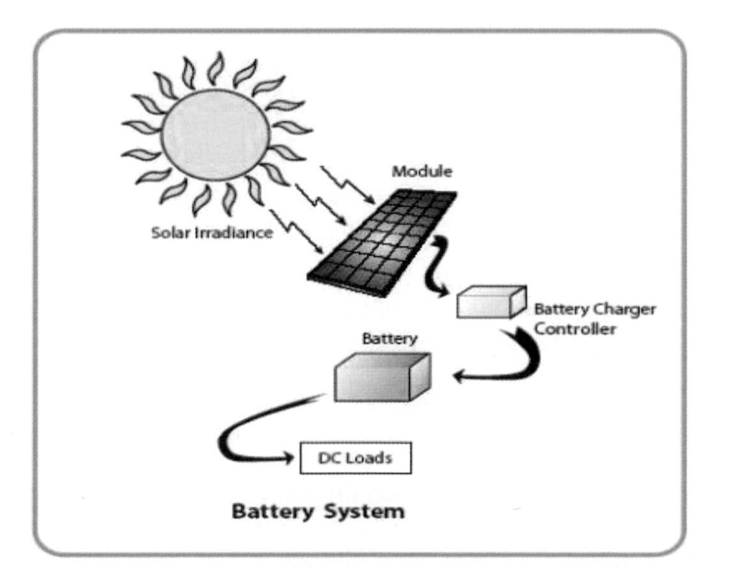

Figure 1.16 (a) Stand alone system.

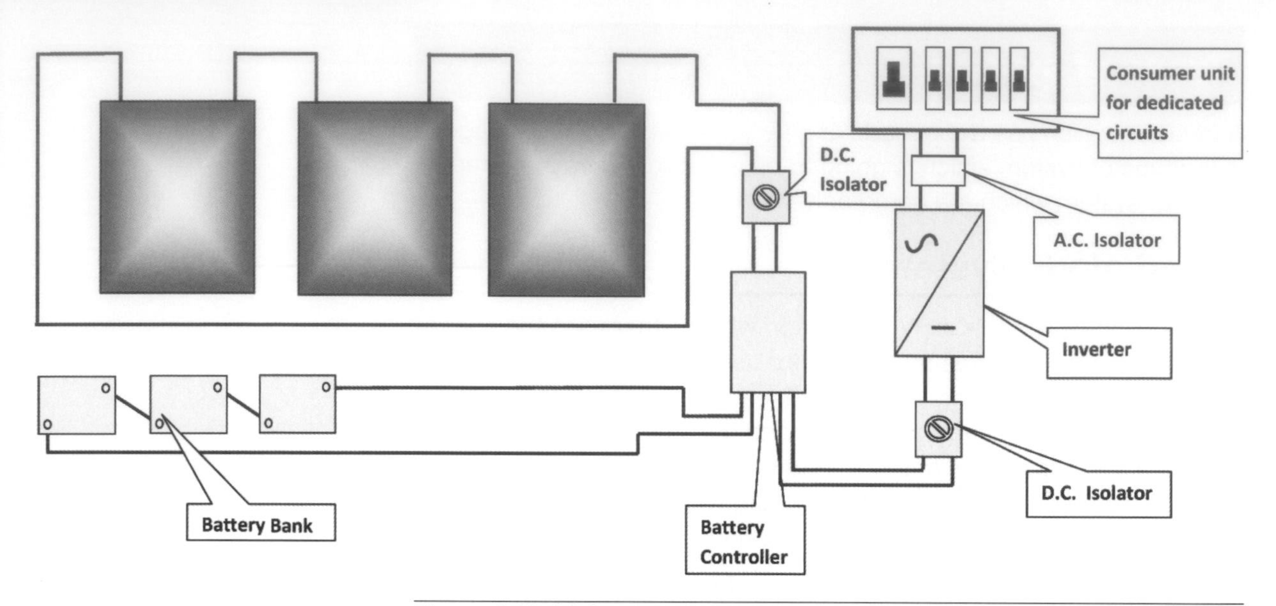

Figure 1.16 (b) Circuit diagram showing how a stand alone system could be used to provide an AC supply.

For instance, it would be necessary for the designer to calculate the daily energy requirements.

Example:

2 x 60 watt lamps for 4 hrs per day

$$120 \times 4 = 480 \text{ watts.}$$

A freezer rated at 165 watts for 10 hours per day (the freezer would not run continually)

$$22 \times 10 = 220 \text{ watts.}$$

A fire protection system rated at 5 watts

$$5 \times 24 = 120 \text{ watts.}$$

The required total energy would be:

$$480 + 220 + 120 = 820 \text{ watts.}$$

As the capacity of a storage battery is rated in amperes we must convert the daily requirements to amperes by dividing the power in watts by the storage voltage

$$\frac{820}{12} = 68.3 \text{ amperes.}$$

This shows us that we would need 68.3 amperes at 12 volts over a period of 24 hours. This is all very well but we are assuming that our PV system is capable of providing enough energy consistently over a 24 hour period to keep our batteries fully charged. As we have seen

Figure 1.17 It is important to use the correct batteries.
Source: Trojan Battery Company.

previously we cannot guarantee that our PV system will be working to its maximum every day, for that reason we should calculate what energy we would require over a 5 day period as a minimum and arrange our energy storage to suit.

Therefore 5 × 68.3 = 314 amperes, this would be the battery storage required to supply the listed items for 5 days.

Great care must be taken when choosing batteries for this type of installation as they are very expensive. There are two types of batteries which are generally suitable for use with PV systems, they are absorbed glass matt and Gel batteries (Fig. 1.17).

Batteries are rated according to their discharge cycles: shallow cycle batteries and deep cycle batteries are readily available. Shallow cycle batteries are the type which would be used to start a car engine. In other words they are designed to deliver a very high current (600/700 amps) for a few seconds, therefore using 15 to 20% of the battery's total capacity. Once the engine is running the battery is then charged up very quickly using the cars alternator.

Providing these batteries are not discharged more than 20 to 25% they will be capable of lasting a very long time, but this type of battery is not well suited to PV installations as on occasions the discharge will be much higher than 25%.

Deep cycle batteries are used to provide a small current for many hours between charges, possibly using up to 80% of the battery's total, and they are also capable of being recharged many hundreds and often thousands of times.

The storage capacity of a battery is given in amp hours (Ah); this is the amount of current which a battery could deliver over a one hour period. A battery rated at 200Ah would be capable of delivering 200 amps for 1 hour, 20 amps for 10 hours or 2 amps for 100 hours.

In the example used earlier we calculated that the required current for 5 days use was 314 amps. For this installation the requirements would be for a battery which had a capacity of at least 314Ah. As with any system which stores energy there will be losses; these losses could be as high as 20% and they must be taken into account when installing the battery storage:

$$\frac{314 \times 100}{20} = 62.8.$$

The battery requirements are now 314 + 62.8 = 376.8Ah.

If we were to use this capacity battery it would be completely discharged after 5 days, but this would not be very good for the battery even if it was a deep discharge type. Bearing in mind that the less a battery is discharged, the more charge cycles it would have and the longer it would last we need to do our best to ensure that the battery is never completely discharged.

Example:

If we had a 200 Ah battery which was rated at 500 discharge cycles with an 80% discharge, we could use 160A of its capacity and recharge it 500 times. If we were to discharge the same battery using only 80A of its capacity we could recharge it 1000 times and the battery would last twice as long.

For this reason it is recommended that a battery in a PV system is only discharged to 50% of its total capacity, this would mean that in our installation with requirements of 314Ah we would need to increase the capacity of the battery storage to 628Ah.

As batteries are not manufactured in infinite sizes we would need to choose batteries to suit our needs as closely as possible, usually larger than required.

Gel batteries are normally rated from 80Ah to 300Ah, and for our installation three 250Ah batteries would be suitable and would last a very long time.

We have seen that with any kind of rechargeable battery storage it is important to ensure that the batteries are never completely discharged; it is equally important to ensure that the batteries are not over charged as this will also reduce the life span of a battery considerably.

To ensure that the batteries are kept in a healthy state a battery charge controller is used (Fig. 1.18).

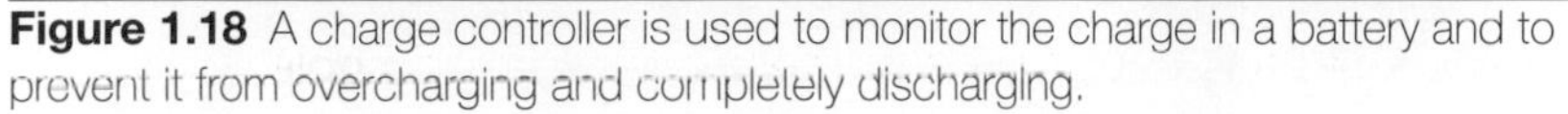

Figure 1.18 A charge controller is used to monitor the charge in a battery and to prevent it from overcharging and completely discharging.

This is a device which monitors the state of the battery system. Once the batteries have discharged to a preset level the charge controller will switch the system off to prevent them being completely drained. This is to prevent irreversible damage to the batteries. The charge controller will also prevent the batteries from being over charged as this could also cause irreversible damage to the battery system.

As we have seen, in its most basic form a stand alone system will be fine to supply d.c. equipment. This is not always practical as usually we have a.c. equipment which we may want to make use of. In many cases a stand alone system may be used to provide a supply for an office, or perhaps a workshop which is remote from a main supply.

This type of installation is ideal for use with PV which has battery storage because the type of equipment used would normally be low current using equipment such as a computer or a lighting system.

Where this is the case we need to convert the generated and stored direct current into alternating current, and for this we need a device called an inverter. One end of it is connected to the battery bank which is d.c and out of the other end we get a.c.

This sort of device is often used in motor vehicles where a piece of a.c equipment is needed, it just plugs into the 12 volt d.c socket which used to be called the cigarette lighter and out of the other end we get 230v a.c. (Fig. 1.19).

Grid connected system

A grid connected system is a PV system which is connected directly to the electrical installation of a building. The PV modules

Figure 1.19 This device will convert 12v d.c. into 230v a.c.

are connected together to form a PV array to suit the proposed installation, it is then connected so that the energy which is generated by the PV system is fed into the installation. This energy can then be used to supply any current using equipment which may be in operation at the time, with any surplus energy being fed back into the national grid (Fig. 1.20).

A major benefit with this type of system is that all of the energy produced by the PV installation will be eligible for a generation tariff, known as a feed in tariff (FIT). This tariff is where the user of the electricity signs an agreement with the DNO (distribution network operator), who will then pay the user a fixed amount of money per unit for any electricity which is generated by the PV installation. This tariff is collectable by the user for all of the electricity generated, even if it is used to provide energy for equipment which is supplied by the electrical installation.

To enable the amount of energy which is being generated to be measured a generation metre must be fitted between the PV inverter and the point of the connection to the fixed installation. The generation metre is usually supplied and installed as part of the PV installation (Fig. 1.21).

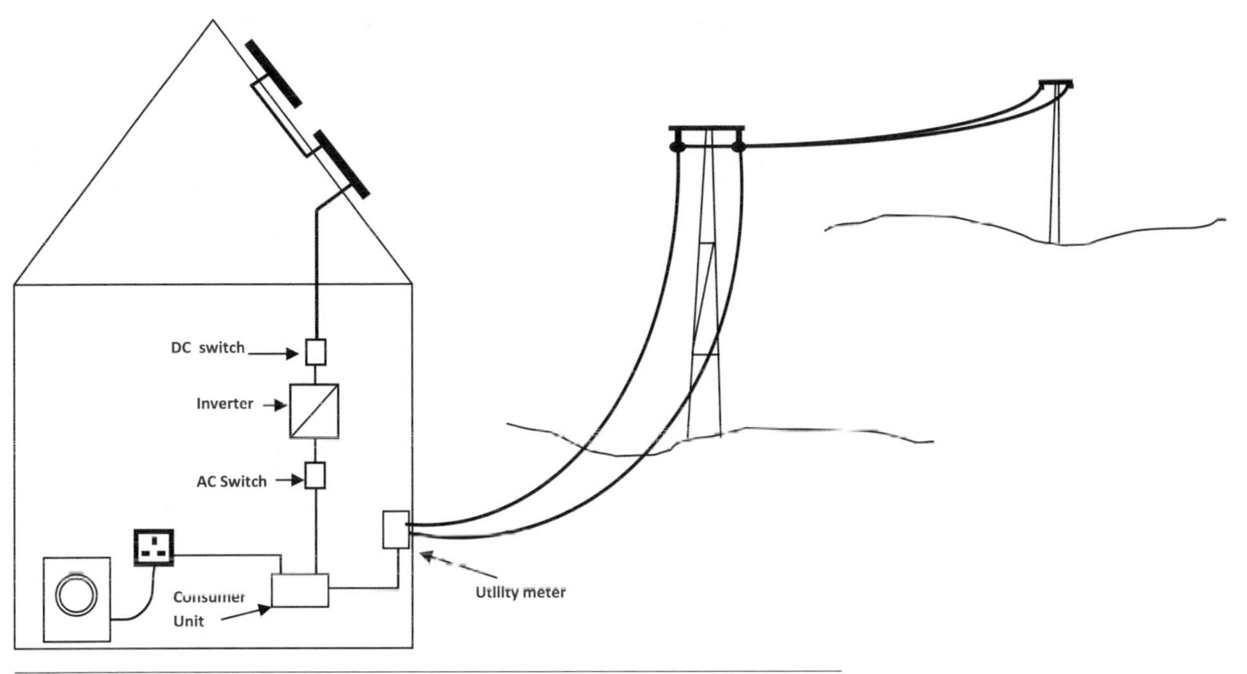

Figure 1.20 Grid connected system

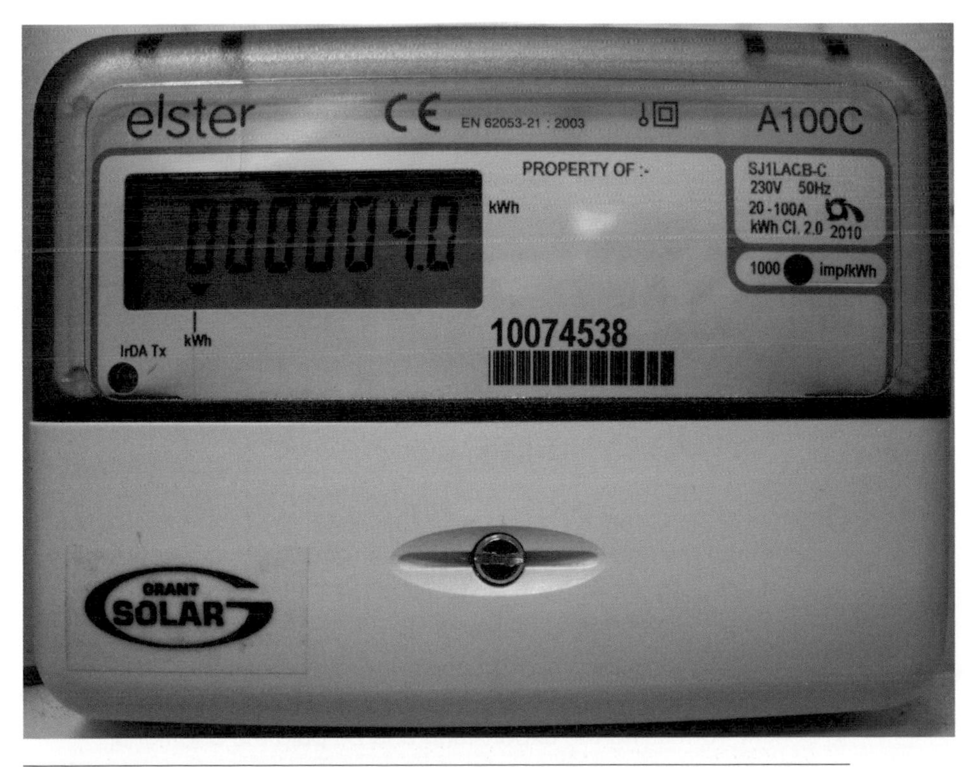

Figure 1.21 A generation metre records the energy generated by the PV system.

The feed in tariffs are set at various levels depending on the maximum amount of energy which the PV installation can produce. At the moment the tariffs levels per unit (kW/h) are:

Up to 4kW new	36.1p
Up to 4kW retro fit	41.3p
Between 4 – 10kW new or retrofit	36.1p
Between 10 – 100kW new or retrofit	31.4p
Between 100kW – 5MW	29.3p
Stand alone systems	29.3p

A new installation would be where the PV has been installed as part of a new building and a retrofit installation would be one that has been fitted onto an existing building. In the table a stand alone system is a system which has been installed purely to feed into the grid and is not connected to an electrical installation.

Most domestic installations will fall into the 4kW category which varies slightly depending on whether the installation is installed onto an existing building, this would be termed as retrofit, or whether the installation was installed onto a new building. The tariff paid for a retrofit system is greater than that paid for new build.

There is another type of tariff which is collectable by the owner of a PV system (as well as the generation tariff which is paid for any energy which is produced), an export tariff. This is an additional tariff which is collectable for any energy which is not used, but is fed back into the grid system for use by the DNO. At the present time the export tariffs for a 4 kW installation are:

Type of installation	Generation tariff (pence)	Export tariff (pence)
Retrofit	41.3	44.3
New build	36.1	39.1

From the table above it can be seen that the export tariff is 3 pence greater than the generation tariff, so 3 pence is the base level which can be renegotiated with the DNO if it is felt that a large amount of energy is being put back into the grid.

All electrical installations are fitted with an import metre which records the amount of electricity being used within an installation. If we want to benefit from using the export tariff and be paid the additional sum for any energy which we do not use we must have a means of measuring it, but most import metres are not capable of doing this.

Over the next few years all properties are going to be fitted with 'smart metres' (Fig. 1.22).

These metres will be capable of measuring energy which is imported and exported and will be able to be read remotely. For installations which do not have import metres which can read exported energy as well or those which do not have smart metres it has been agreed with all of the DNO that an export tariff will be paid on 50% of any electricity which is generated.

Figure 1.22 This smart metre will record the amount of energy fed back into the grid as well as the energy generated by the PV system.
Source: Shutterstock.

As we have seen, photovoltaic modules produce direct current, and as we know our main electricity supply system is alternating current. For the system to work we have to convert the direct current into alternating current, so once again we need an inverter (Fig. 1.23).

The type of inverter required for connection to the grid system is much different than the inverter which we would use in a stand alone system.

It is the job of the inverter not only to change the direct current into alternating current but to synchronise the frequency of the converted alternating current to that of the supply. Because the inverter is being

Figure 1.23 An inverter will convert the generated direct current into alternating current and then synchronise it with the frequency of the supply.

connected to the grid system the regulatory requirements are very precise.

Engineering recommendation G83/1 is a document which sets out the requirements for generation systems up to 16A. For safety reasons all inverters which are used to connect to the grid must comply with all of the requirements of this document.

The inverter must have a certificate of compliance (Fig. 1.24), and this will be supplied with the inverter.

Grid connected inverters must also have an inbuilt protection system which shuts down the device completely in the event of a loss of the supply from the grid. Imagine the danger if the network operator had to carry out some repairs or maintenance on the electricity supply system. They could turn off the supply system, but it could still be energised by any number of microgeneration systems which were still operating and connected to the grid via their inverters. This type of situation is termed 'islanding' and must be avoided.

The requirements which are set out for inverters are that they must shut down when:

- there is a loss of main supply
- the operating voltage rises above 264 volts between line and neutral
- the operating voltage falls below 207 volts between line and neutral
- the operating frequency rises above 50.5Hz
- the operating frequency falls below 47Hz.

The connection of the PV system to the main installation can be quite a simple process as all that is required is for the supply (outgoing) side of the inverter to be connected directly to a spare way in a consumer's unit (Fig. 1.25).

Where this method of connection is used it is advisable to carry out a periodic inspection on the existing installation, this is to ensure that the existing installation is safe to connect to.

Another method, and the one which I prefer, is to terminate the PV system into a one-way consumer unit (Fig. 1.26) and then connect the consumer unit tails by fitting a mains block (Henley Block) between the existing metre and the main switch (Fig. 1.27).

By using this method it is not necessary to touch the existing consumer's side of the installation and the certification required is only that which will need to be completed for the PV installation.

In most cases the inverter will be class 2 rated and will not need earthing, although it is always necessary to check for the class 2 symbol (Fig. 1.28).

Bureau Veritas Consumer Product Services GmbH
Businesspark A96
86842 Türkheim
Germany
+ 49 (0) 8245 96810-0
cps-tuerkheim@de.bureauveritas.com

Certificate of compliance

Applicant: **SolarEdge Technologies Ltd.**
Abba Eban 1a
Hertzlia 46725
Israel

Product: **Automatic disconnection device between a generator and the public low-voltage grid**

Model: **SE3300; SE4000; SE5000; SE6000**

Use in accordance with regulations:

Automatic disconnection device with single-phase mains surveillance in accordance with Engineering Recommendation G83/1 for photovoltaic systems with a single-phase parallel coupling via an inverter in the public mains supply. The automatic disconnection device is an integral part of the aforementioned inverter. This serves as a replacement for the disconnection device with isolating function that can access the distribution network provider at any time.

Applied rules and standards :

Engineering Recommendation G83/1.

The safety concept of an aforementioned representative product corresponds at the time of issue of this certificate of valid safety specifications for the specified use in accordance with regulations.

Report number: **09TH0001-G83**
Certificate nummer: **U10-409**
Date of issue: **2010-09-17** **Valid until:** **2013-09-17**

Achim Hänchen
CERTIFICATION

Figure 1.24 All inverters must have a certificate of compliance.

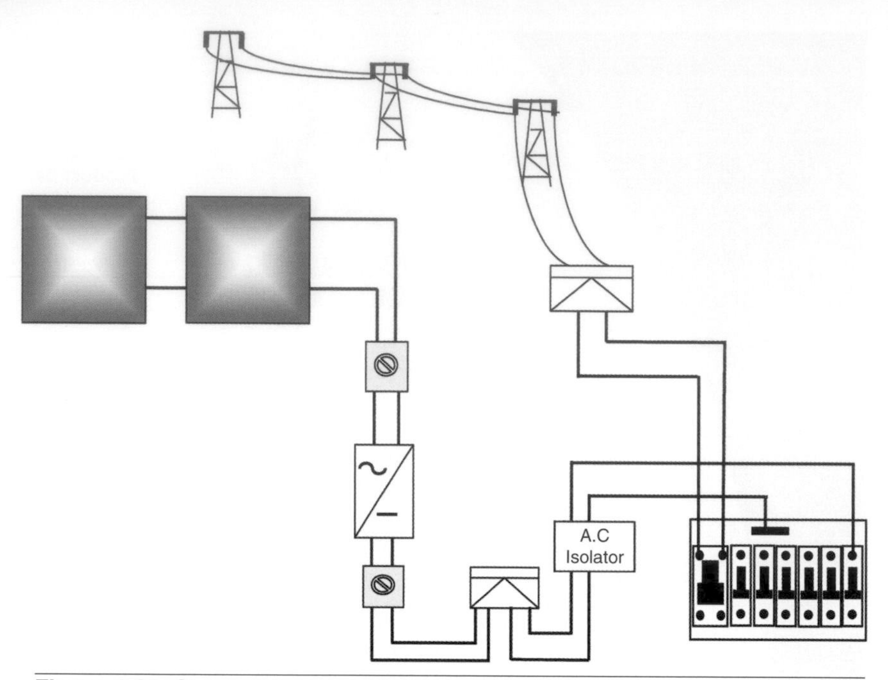

Figure 1.25 Circuit diagram for a PV system connected to a spare way in a consumer's unit.

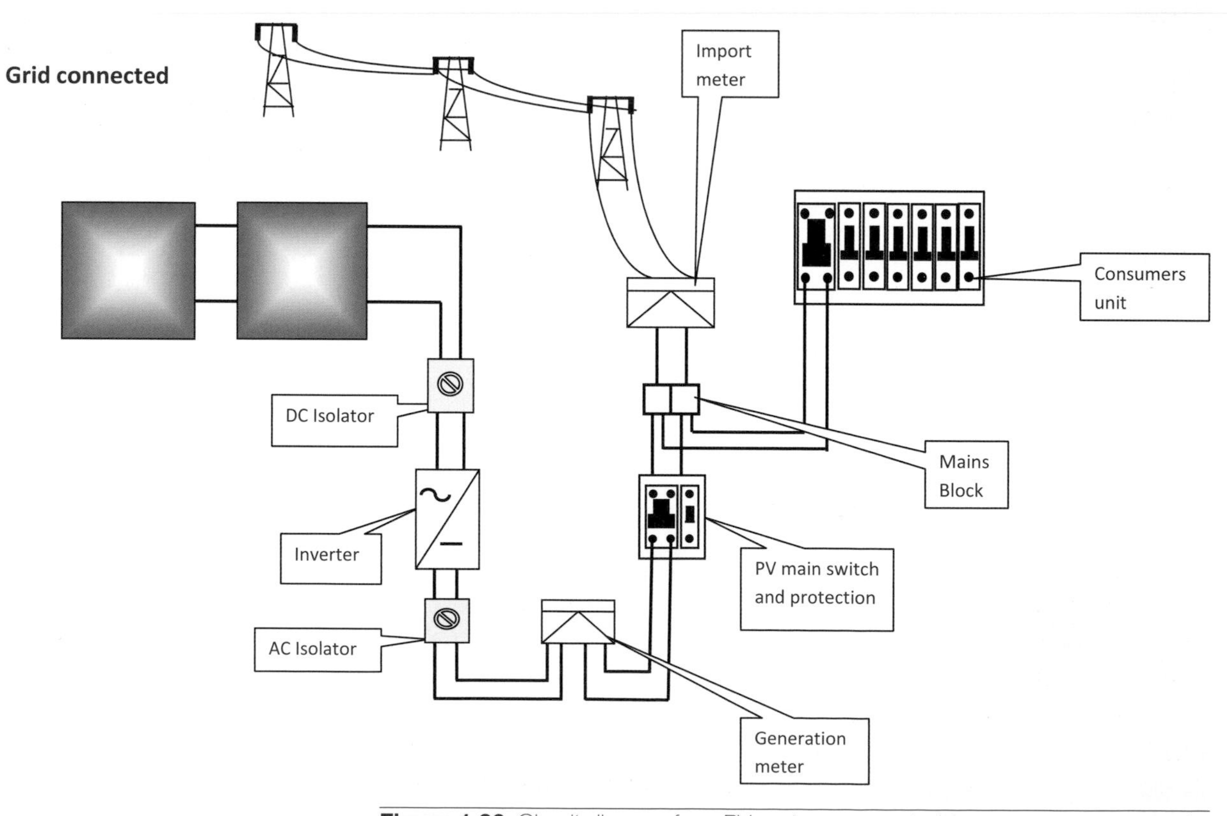

Figure 1.26 Circuit diagram for a PV system connected to a one-way consumer's unit.

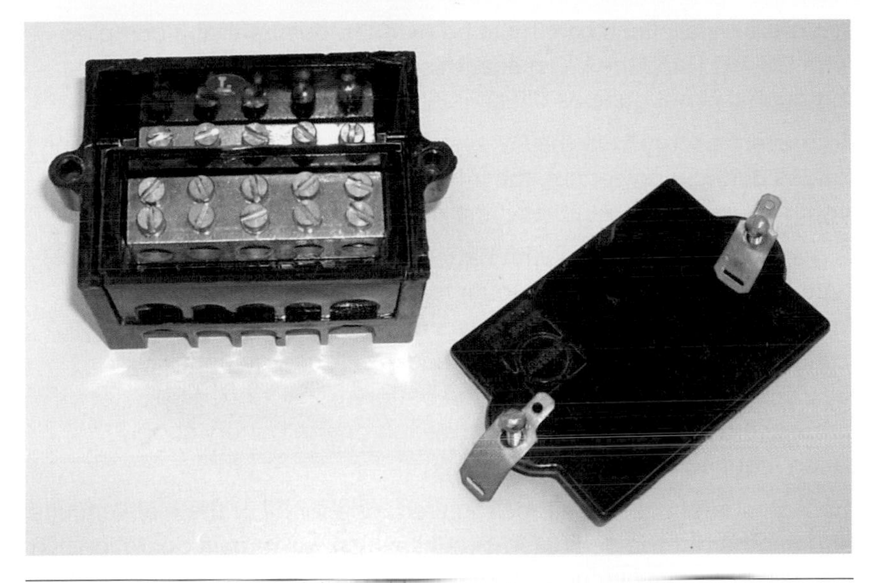

Figure 1.27 A Henley Block is used for connecting to the metre tails.

Most inverters also contain an isolating transformer which of course eliminates the requirement for an earth on the load side, although as in most cases the inverter is connected to an installation which is earthed, the earthing system will provide fault protection for any circuits which are connected to the consumer's unit as shown in Figures 1.20 and 1.26.

Figure 1.28 Unless equipment has this symbol it must be assumed that it is class 1.

Both of these methods will allow any generated energy to be used as required and any that is not used to be fed back into the grid. The connections which are required from the inverter to the consumer's unit must be carried out by a competent part P registered installer and certificates must be provided to document the installation details.

A metre will have to be installed as shown in Figure 1.21 and this will record any energy generated by the system.

Grid tie system with battery backup

With this type of system the user of the installation has the best of both worlds as it combines a grid connected system and a battery system. This is very useful in areas where power cuts are common or where a supply to some dedicated circuits must be maintained if there was a loss of supply.

This type of system cannot be connected directly to the consumer's unit in the normal way, as we have seen in the previous chapter it is important to ensure that the supply cables do not remain live in the event of a power cut.

As for all installations care must be taken to ensure that it complies with BS 7671 and the DTI guide, this involves a bit of thought although it is not quite as difficult as it may seem.

To prevent energy from the PV system being fed back into the supply cables during a power cut, the inverter must isolate the PV system from the main supply cables if the main supply is lost.

A grid connected system with battery backup would need a special type of inverter with two a.c. outlets, one of the outlets would immediately shut down if there was a power cut and the other outlet would switch to the battery backup. A charge controller would also be required to keep the battery bank topped up, prevent it from over charging and switch on the battery supply when required.

Two consumer units would be required, one for all of the main circuits and another for the dedicated circuits which will remain operational in the event of a power cut.

The main consumer's unit would be fed from the main supply and would also be connected to the inverter, a consumer unit containing the dedicated circuits would be connected to the main supply via the inverter. When the main supply fails the inverter and the battery controller will switch to battery operation only and isolate the main supply until it was reinstated.

The control circuit for the contactor and the battery controller would need to be connected to the main consumer's unit (Fig. 1.29).

This type of system may require the installation of two generation metres as shown in the drawing, although some inverters will incorporate a generation metre. Other wiring arrangements may also be used provided they comply with all of the regulations.

To ensure that the system was completely isolated from the supply in the event of a power failure, the system should not make use of any earth provided by the supply DNO and the installation should use a TT system.

Grid fallback system

On small installations the grid tie system could be altered slightly to provide what is known as a grid fallback system. In this system the energy required by dedicated circuits is supplied via an inverter which is connected to the battery storage system and the PV array. Energy from the grid would only be needed if the batteries became discharged, the grid would then be used to supply any energy required until the batteries were recharged by the PV array (Fig. 1.30).

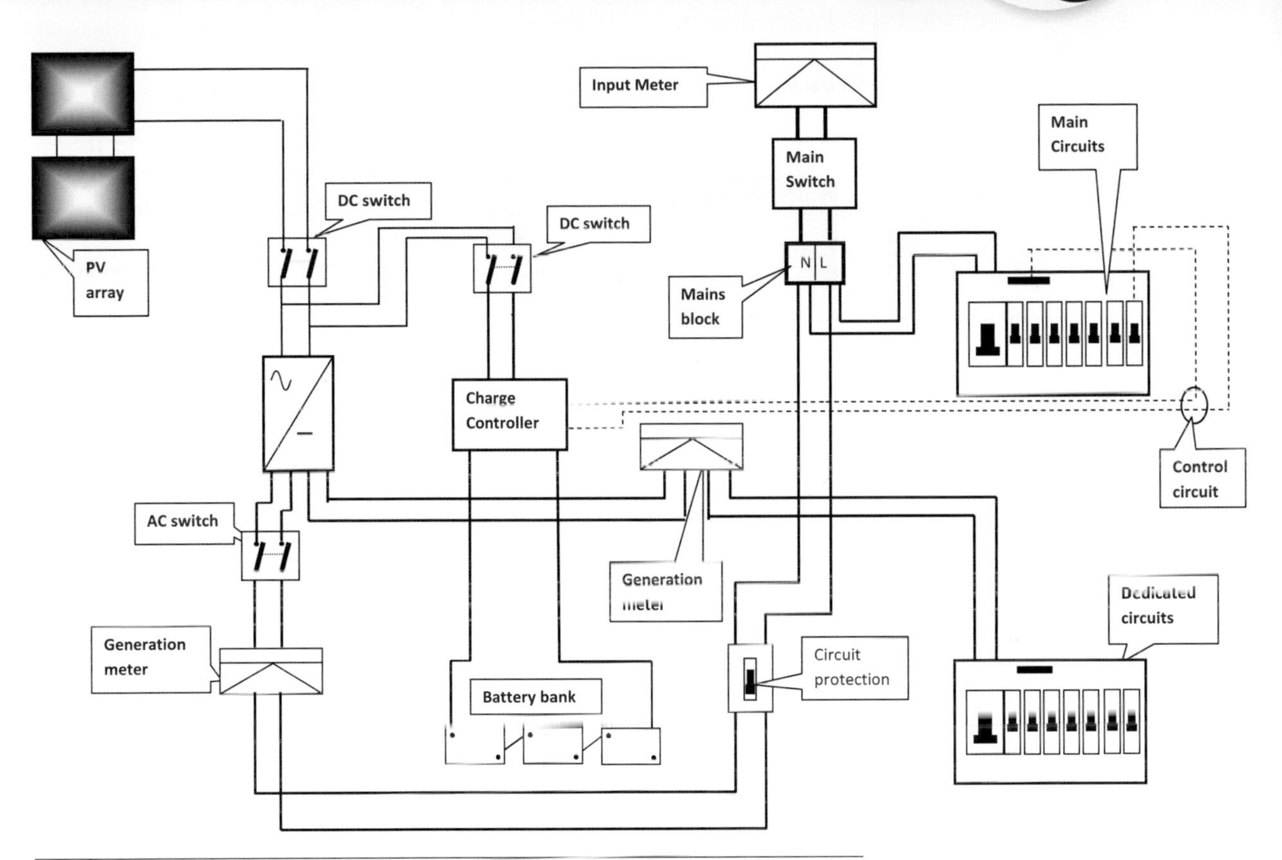

Figure 1.29 Circuit diagram for a grid tie system with battery backup.

Installation requirements

Having calculated how many photovoltaic modules that can be installed it is very important to get the positioning of them correct. As we have seen, facing south is the best direction and if the angle is to be fixed, which most usually are, an angle (tilt) of between 30 and 40 degrees is the most suitable (Fig. 1.31).

The site survey must include looking to see if any shadows are likely to be cast across the array, as shading could be caused by chimneys, dormer windows or even trees and bushes. Whatever the cause, direct shading must be avoided, as a shadow cast on any part of the PV array will result in a massive loss of efficiency, possibly up to 80%. A shadow could also cause damage to the module which it falls onto, this is due to the module becoming a load rather than a generator.

Figures 1.32 and 1.33 show how a shadow could be easily overlooked when carrying out a survey. Figure 1.32 shows the roof

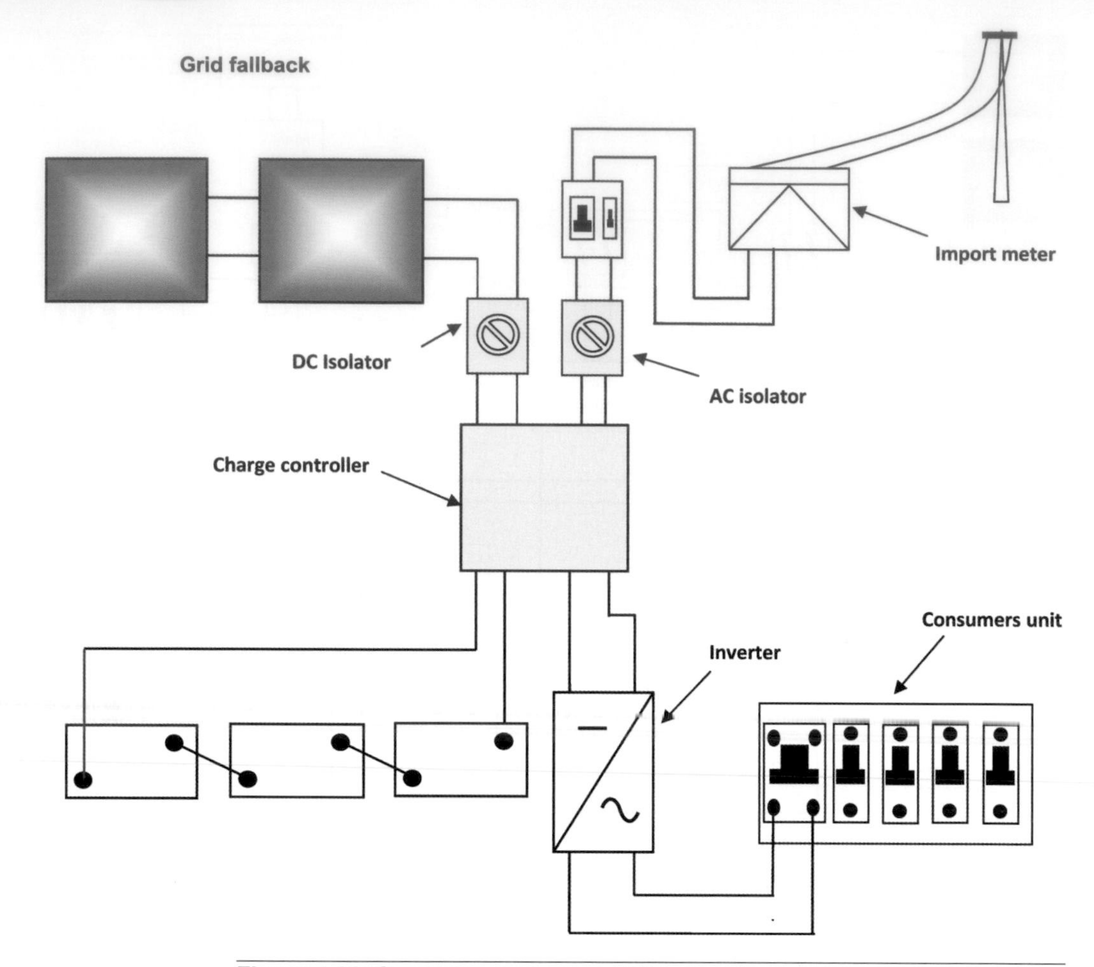

Figure 1.30 Circuit diagram for a grid fallback system.

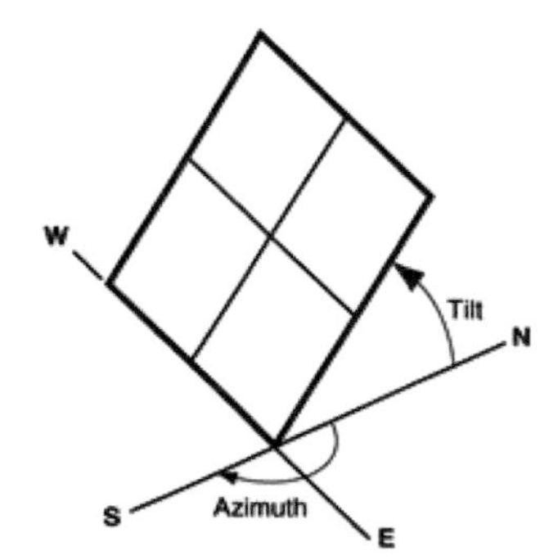

Figure 1.31 If a tracking device is not used then a panel facing south at an angle of between 30 and 40 degrees will give the best results.

on a cloudy day and no shadows are being cast by the sun, however Figure 1.33 tells a completely different story as the sun is now shining and casting a shadow right across the roof.

It would not be worth considering installing a PV installation on this roof as it would generate very little for a large part of the day.

Good quality modules will have blocking diodes fitted as standard which will prevent damage but will not prevent the loss of energy.

In most instances planning permission is not required for the installation of PV systems, but information on the planning requirements for England can be found in the statutory document entitled Town and County Planning, England order 2008 No. 675. The only planning requirements for the majority of installations are that the panels do not protrude above the roof line, and that they do not sit more than 200mm above the slates or tiles.

Figure 1.32 There will be no shadow if the sun is covered by a cloud.

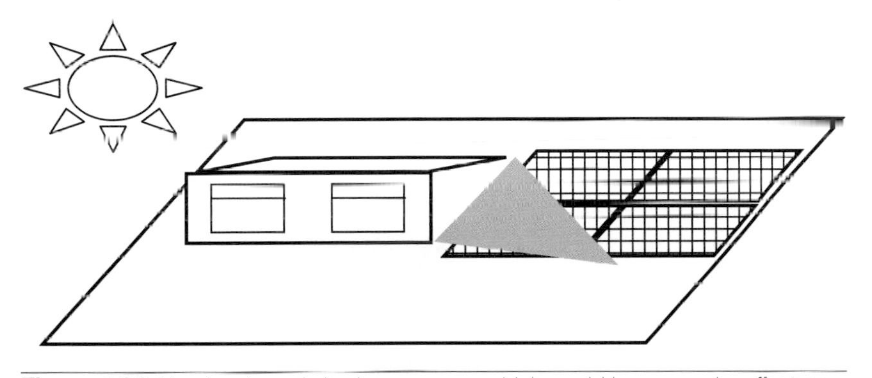

Figure 1.33 No clouds and shadows appear which could have a major effect on the amount of energy produced by the solar array.

Consideration should also be given to additional load on the roof as the array will be quite heavy, it will also change the characteristics of the roof with regards to wind load as well. It is a good idea to have a structural survey carried out to ensure that the roof is strong enough to cope with the added loads. It may be that a small amount of structural work is required before the array is fitted.

Before attempting to install the PV modules onto a roof it is extremely important that all safety precautions are taken. One of the most obvious of course is to have a suitable scaffold erected and also to ensure that there is enough manpower available when the modules have to be lifted onto the roof, as they are very heavy.

It goes without saying that the array must be fixed securely to any roof and there are various fixings manufactured for all types of roof as tiles and slates come in all different shapes and sizes. The fixing methods for PV panels are the same as for solar thermal panels.

Figure 1.34 A rafter fixing.

Usually a bracket is slid under the tile and screwed to the roof rafter. This bracket will protrude from between the tiles and will then have a channel fixed to it which in turn will have the panel fixed to it. Figure 1.34 shows how one type of rafter fixing would be used.

Once the brackets have been securely fixed to the rafters and the tiles or slates have been replaced a channel is fitted to the brackets (Fig. 1.35).

Before placing the panels on the roof it is important that each panel is tested for correct operation. This is a simple process and is really

Figure 1.35 Rails fitted to brackets with some panels.

just a matter of measuring V_{oc} and I_{sc}. If all of the panels are giving the same or very similar readings it stands to reason that they must be working correctly.

Once the panels have been fixed to the roof they need to be electrically connected to form strings, the cables from the stings will then be passed through the tiles or slates into the roof space. Care must be taken where the cables pass between the tiles as tiles can have sharp edges which could cause damage, tight bending radiuses should also be avoided. One method of avoiding this is to pass the cables through a vent tile which could be positioned particularly for the purpose. Another method would be to drill through a tile or a slate and then make a lead slate to fit over it which would allow the cables to pass through it (Fig. 1.36).

Once the cables are in the roof space each string should be tested for its open circuit voltage and its short circuit current before connecting them together.

This is to ensure that the strings are operating as they should be.

Where there are four or more strings each string will need to be fused and any fuses used must be d.c. rated as they have completely different characteristics than a.c. fuses.

Figure 1.36 Where the cables penetrate the roof care must be taken to ensure that they will not be damaged over a long period of time, the roof must also remain watertight.

Figure 1.37 Isolators used on the d.c. side of the installation must be d.c. rated. If they are not then the contacts will be damaged when the installation is isolated under load.

It is important to remember that each PV module will generate electricity as soon as it is exposed to any kind of light, this of course means that the cables will be live and will not be able to be isolated.

After confirmation that each string is operating as expected, they may be joined together to form the completed array. Although it is not a requirement it is a good idea to provide an isolator for each string, these isolators must be d.c. rated (Fig. 1.37).

This will slightly increase the cost of the installation but it will allow each string to be isolated individually, whilst allowing the other strings to continue operation should any type of maintenance be required. The connection of the strings must be made with extreme care because loose connections can cause arcing which will increase the risk of fire. Because of the nature of the current, d.c. causes far greater arcs than you would expect from an a.c. circuit.

Special touch proof plug connectors are available which form very low resistance connections (Fig. 1.38). These connectors are male and female and are usually made on to the ends of the cables as

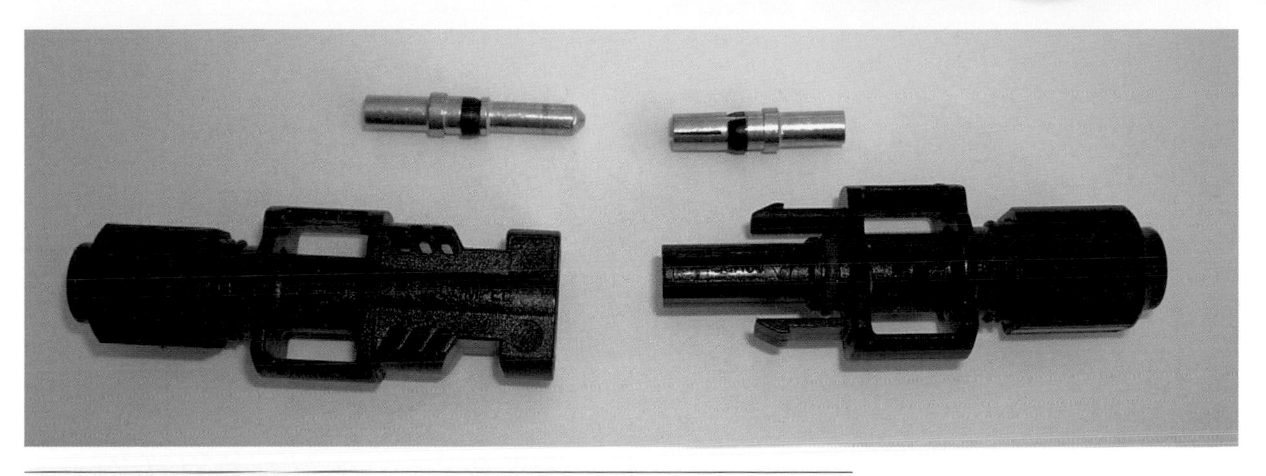

Figure 1.38 PV connectors must be used, these are touch proof and will also be suitable for use outdoors.

required. The inserts are fixed to the ends of the cables before inserting them into the touch proof connectors.

For installations which call for the joining of two cables into one, branch connectors are available (Fig. 1.39) and the use of these can often save a considerable amount of cable.

It is usual to make the final connections on site. This requires the use of a special crimping tool which fixes the male and female ends to the conductors.

The first step is to strip the cable back to the required length (Fig. 1.40).

The connectors are either male or female (Figs. 1.41 and 1.42).

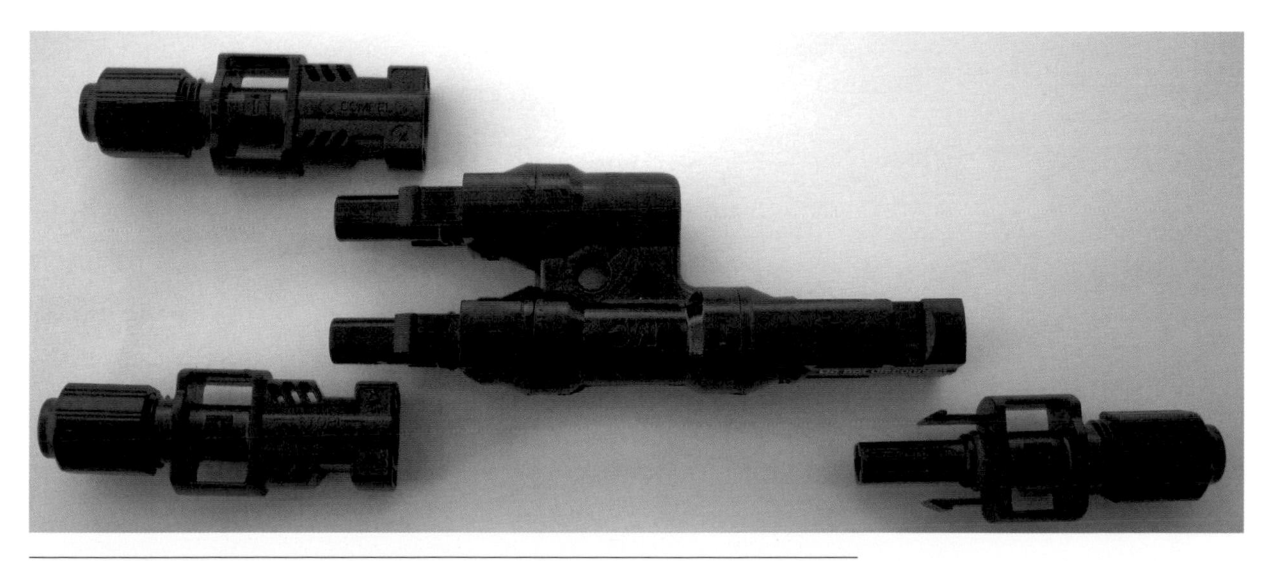

Figure 1.39 Where two strings need to be joined, branch connectors can be used.

Figure 1.40 Cable stripped to the correct length.

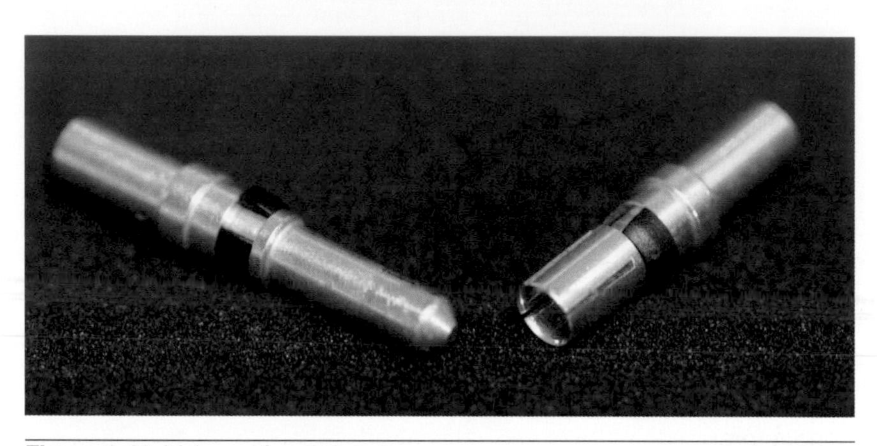

Figure 1.41 Male and female inserts.

Then slide the female or male connector on to the bare end of the cable (Fig. 1.42a) and use the crimping tool to fix the connector securely.

(a)

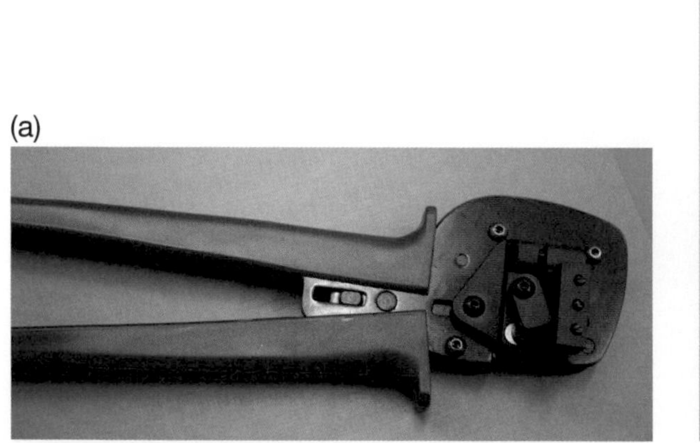

(b)

Figure 1.42 Cable being crimped using the correct type of crimping tool.

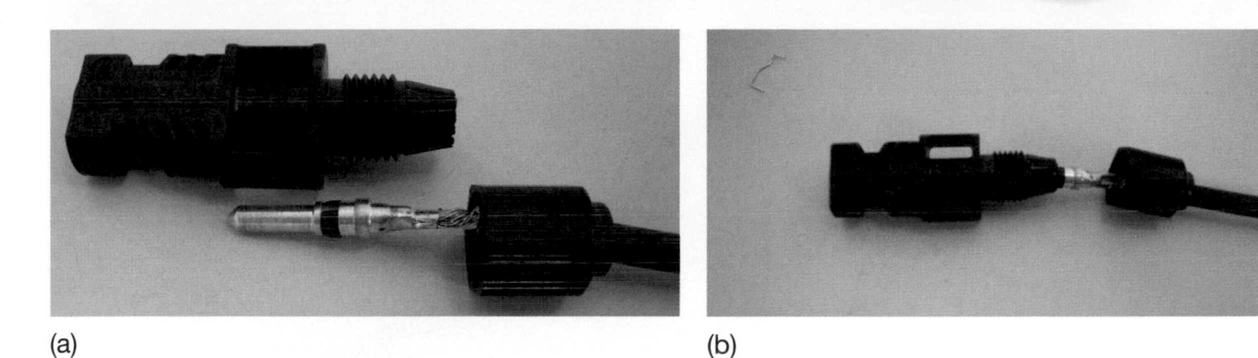

(a) (b)

Figure 1.43 A touch proof connector.

Once the insert has been correctly crimped it can then be inserted into the touch proof connector (Fig. 1.43a/b and 1.44).

When the strings have been connected to form an array the array must be connected to an inverter. It is acceptable if required to install the inverter for the system in the roof, close to the array, although in most installations it is more practical to install the inverter near the consumers unit for ease of access. The cable used to connect the array to the inverter must be d.c. rated and marked to show that it is a d.c. cable with each conductor's polarity clearly indicated.

In most cases the simplest way of connecting the PV array to the inverter is to use purpose designed single conductor PV cables, these are H07RNF cables (Fig. 1.45).

Other types of cable may be used if required, for instance in a particularly harsh environment it may be that single core steel wire armour cable would be required

Standard singles cables can be used providing they are mechanically protected either by plastic conduit or, if the environment is likely to be harsh, then earthed metal trunking should be used. Each conductor should be in its own conduit.

Figure 1.44 Completed female connector.

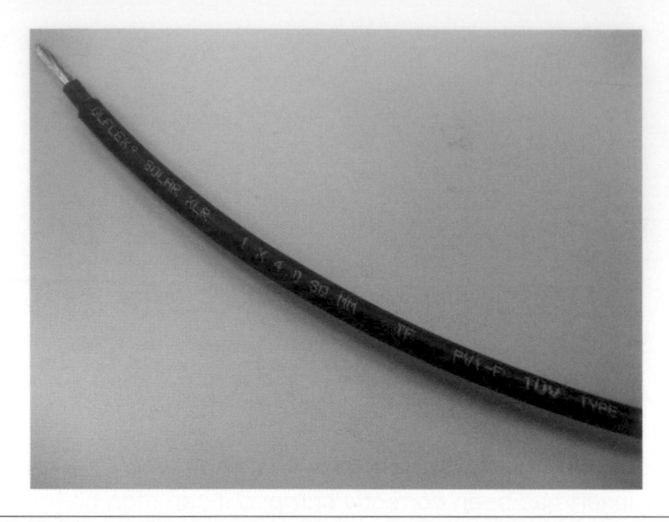

Figure 1.45 It is usually easier and better to use PV rated cables.

The cable should be routed to avoid any possibility of mechanical damage or damage which could be caused by fauna. As mentioned, any damage to the cable would greatly increase the risk of fire as direct current causes far greater arcs than alternating current. Before the cable enters the inverter a means of isolation must be provided to enable the PV array to be completely isolated from the inverter. The isolator must be d.c. rated clearly labelled and lockable (Fig. 1.46).

D.c. rated cables must also be used to connect the inverter to the d.c. isolator.

Figure 1.46 It is very important to ensure that the isolators are correctly labelled.

Sizing of the cables from the array to the inverter is very important, as with any other circuit the cable has to be able to carry the maximum current at the supply voltage without damage.

In a PV system the current is the maximum which the array can produce. This is known as the short circuit current (I_{sc}) and the voltage is the maximum voltage which could be produced which is known as the open circuit voltage (V_{oc}). The rules are that the cables on the d.c. side must be able to carry 1.25 times the maximum current (I_{sc}) and 1.15 times the maximum voltage (V_{oc}).

Of course another consideration is the permitted voltage drop, it is really important that this is kept to a minimum in the d.c and a.c. parts of the installation. A loss in voltage will result in a loss of power and of course the main reason for having a PV installation is to save energy. For this reason it is not a bad idea to oversize cables slightly as larger cables have less resistance which in turn will result in a smaller voltage drop.

As an example let us assume that we have three strings each producing a current of 7.4A at a voltage of 104.4V. It would be easier in most cases to connect all of the strings together using a connector box sited in the most convenient position, usually it would be sited to ensure that the cable runs from the strings to the connector box are not too long (Fig. 1.47).

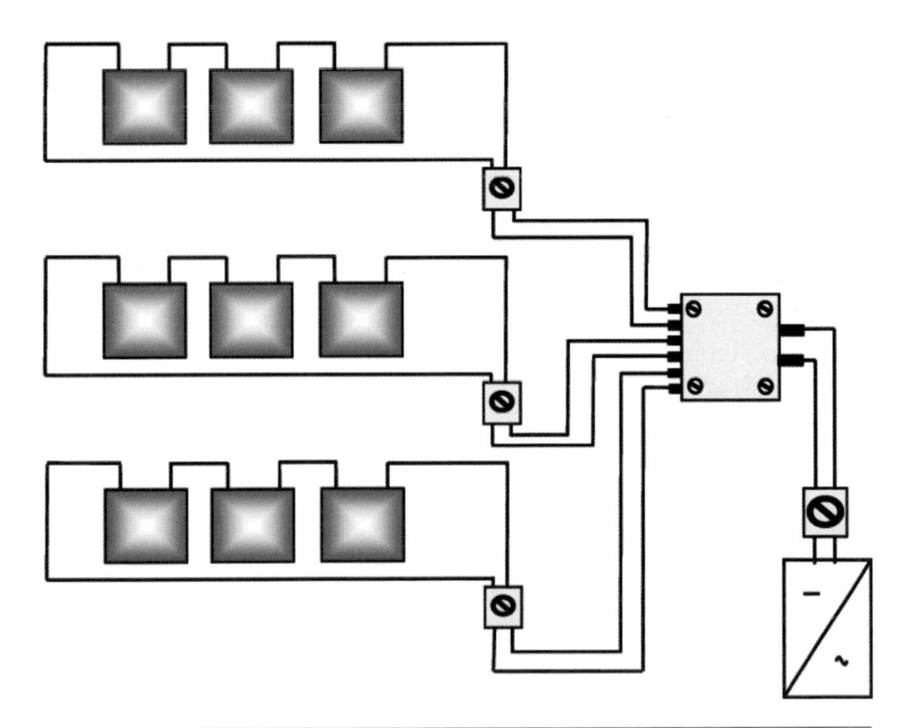

Figure 1.47 Connector boxes should be connected as close to the PV arrays as possible, this will help to reduce voltage drop.

If the cables from the each string were 3 metres long (*maximum, as in practise they will not all be the same length*), with a cross sectional area of 1.5mm^2, the resistance of the cable would be 12.10mΩ/m at a temperature of 20°C.

We must remember that the current will flow in both the positive and negative conductors, so this will result in the resistance being doubled

$$12.10 + 12.10 = 24.20\text{m}\Omega/\text{m}.$$

As the cables are three metres long

$$3 \times 24.20 = 72.6\text{m}\Omega.$$

It is always easier for us to work in ohms, to convert mΩ to Ω we must divide by 1000

$$\frac{72.6}{1000} = 0.0726\Omega.$$

This calculation shows us the maximum resistance of each string cable. The problem is that any cable which is likely to pass behind a PV array must be rated at 80°C and our resistance value is calculated at a temperature of 20°C.

The type of copper which is used in BS cables increases in resistance with a rise in temperature, this increase is 2% for each 5°C rise in temperature.

Although our cables may not reach 80°C we have to take the worst case scenario when working with cables as we need to be certain that any installation is as safe as it can possibly be. As the resistance rise is 2% for each 5°C and the temperature of the conductors could rise from 20°C to 80°C, we can see that the temperature rise will be 60°C.

This now requires a simple calculation to find the percentage rise in temperature

$$\frac{60}{5} \times 2 = 24.$$

The resistance of the cables will rise by 24%. To complete the calculation to find the resistance of the cables by multiplying the resistance value by 1.24

$$0.0726\Omega \text{ x } 1.24 = 0.09\Omega.$$

This is the resistance of the conductors from a string to the connection box.

Now we must calculate the resistance of the conductors from the connection box to the inverter, let's say it is a 2.5mm^2 copper conductor 12 metres long. As each string has a current of 7.4A the conductors must now be able to carry the total current being produced by all three strings which is 3 x 7.4 = 22.2A (Fig. 1.48).

The size of the conductors will now depend on the installation method, but for the sake of the calculation let's say we are going

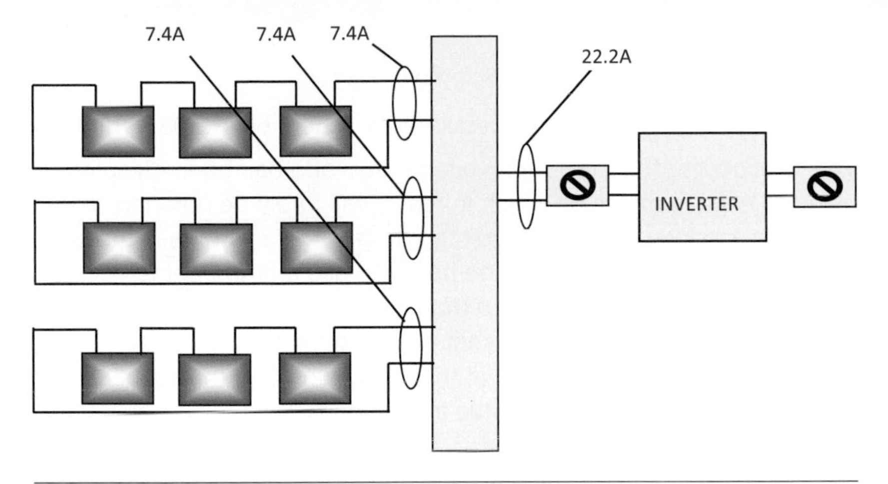

Figure 1.48 Three strings connected to one inverter.

to use a $4mm^2$ copper conductor. Not only will this be able to carry the current easily it should help us to keep our voltage drop to a minimum, although of course we will not know for sure until we carry out the calculation. Another point to remember is that the cable which runs from the connector box to the inverter will not pass behind the PV panels, therefore they can be calculated for a temperature of 70°C not 80°C as before. The multiplier for 70°C is 1.20.

A $4mm^2$ conductor has a resistance of 4.61mΩ/m, again there are two conductors, 4.61 + 4.61 = 9.22mΩ/m.

To find the total resistance of this section of cable we need to carry out the previous calculation but using the figures for this cable

$$\frac{12 \times 9.22 \times 1.20}{1000} = 0.132\Omega.$$

We have now calculated the resistance for each section of the cable at its maximum permitted temperature.

From the string to the connector box is 0.09Ω and from the connector box to the inverter is 0.132Ω.

All we need to do now is carry out a simple ohms law calculation for each section to find the total actual voltage drop from any string to the inverter.

From a string to the connector box the current in the conductors will be 7.4A with a resistance of 0.09Ω:

Resistance x current = volts

$$0.09 \times 7.4 = 0.66\text{v}.$$

From the connector box all of the strings will be connected together, for that reason the current in the conductors will be 22.2A (Fig. 1.48). We have calculated the resistance of this section of cable as 0.338Ω, our calculation is:

Resistance x amps = voltage

$$0.132 \times 22.2 = 2.93.$$

The total voltage drop for the installation will be 0.66 + 2.93 = 3.59v.

This is of course the maximum voltage drop and has been calculated using all maximum values which in reality will rarely be reached. The generally accepted maximum voltage drop for PV systems in the UK is 3%. For this reason it would be better to use a 6mm^2 cable which would have a lower voltage drop than the original choice of 4mm^2. It should always be remembered that even a small loss will become a very large loss over the lifetime of the installation and the installation of a larger cable will cost very little in relation to the cost of the whole installation.

If 6mm^2 has a resistance of 3.08mΩ/m, the calculation would be:

$$\frac{(3.08 + 3.08) \times 12 \times 1.2}{1000} = 0.088\Omega.$$

Resistance × amperes = 0.088 × 22.2 = 1.95v.

This would be much better and would result in far less energy loss.

In installations where the strings are situated apart, possibly on different parts of the roof, the cables from each string can be run separately if required. In these situations the voltage drop must be calculated for each string, using the previous example.

The cables from the string to the connector box would have a voltage drop of 0.66v.

From the connector box to the inverter we will calculate for a 2.5mm^2 cable as the current in each string will be 7.4A.

2.5mm^2 conductors have a resistance of 7.41mΩ/m. Two conductors will have a resistance of 14.42mΩ/m

$$\frac{14.42 \times 12 \times 1.2}{1000} = 0.2\Omega.$$

Resistance × current = voltage

$$0.2 \times 7.4 = 1.48 \text{ volts}.$$

Our choice of 2.5 mm^2 cable will be fine.

Siting of the inverter

The ideal place to put the inverter for the installation is as close to the incoming supply point as conditions allow. However in some cases the inverter can be situated in the roof space near the entry point of the cables from the PV strings. Where the inverter is situated in a roof, access to it must be safe. This would usually require the installation of a loft ladder and boarding in the roof to form a safe walkway.

The main advantage of installing an inverter in a roof is that the cables from the inverter to the connection of the a.c. system would only need to be a.c. rated cables. This of course would save having to use the cable installation method which would be required for the d.c. cables, although of course the a.c. cables would need to be installed to comply with BS 7671 wiring regulations.

Wherever the inverter is situated it is important that the d.c. side of the installation is fed into the inverter via a d.c. rated switch, because when direct current is switched it creates a much larger arc than the same amount of current in an a.c. circuit. Where the strings are run to the inverter separately they must be individually switched before the inverter (see Fig. 1.47).

It must be possible to isolate the a.c side of the inverter. An a.c. isolator with the correct current rating can be used (Fig. 1.49) or, depending on the installation, a double pole circuit breaker or residual current device may be suitable.

It is important that the correct size of inverter is selected, as an inverter which has a rating in excess of the requirements of the installation will result in high power losses. The losses in a correctly selected inverter can amount to as much as 20%, so further losses must be avoided. Where it is intended to extend the PV installation at

Figure 1.49 Correctly labelled isolator on the a.c. side of the inverter.

a later date a larger inverter can be installed but of course the drop in efficiency will be unavoidable until the installation is extended.

An inverter which is too small for the installation will result in the inverter being damaged.

Testing and commissioning

As with any other electrical installation it is a requirement that the installation is inspected, tested and commissioned by a competent person and that the correct certificates are completed and handed to the customer along with the user's instructions provided by manufacturers.

A visual inspection forms a very important part of the commissioning process and is the first action.

Visual

Panels

- PV panels are to British Standards.
- Are the PV panels correctly fixed to the roof or other part of the building?
- Where the panels form part of the roof, is the roof weathered properly with the correct type of flashings installed where required?
- Is the roof suitable for the additional load?

Cables

- Where the cables penetrate the roof the cables are not bent with too tight a radius, and there are no sharp edges which could damage the cables.
- Are the cables in the roof properly supported (clipped)?
- Correct type of cable connectors used.
- Correct type of cables used for d.c. and a.c. installation.
- Cable correctly selected for current rating and voltage drop.
- Has mechanical protection been provided for cables?
- Are the d.c. cables correctly labelled for identification purposes?
- Have string fuses been fitted where four or more strings have been installed?

Control equipment

- Has a d.c. isolator been provided before the inverter on the d.c. part of the installation? This device can be lockable in the off position only.

- Is the inverter correctly rated and manufactured to an appropriate standard?
- Has an a.c. isolator been fitted on the a.c. side of the inverter? This device must be lockable in the off position only.
- Is all equipment correctly labelled?
- If the installation is going to be connected to the grid and the feed in tariff claimed, are all products used compliant to MCS (Microgeneration Certification Scheme) standards?

General

Labelling and identification is required throughout the installation to reduce the risk of accidents being caused due to misidentification, labels would be required for:

All conductors

- On the d.c. side of the installation, are the conductors correctly identified? The positive must be brown and the negative grey. A small length of coloured sleeving is suitable.
- On the a.c. side of the installation, positive is brown and neutral is blue.

Isolation points

- D.c. side of inverter.
- A.c. side of inverter.
- Warning of dual supply, indicating isolation point for PV and mains supply.
- Main isolation a.c. isolation point.

Warning signs

- PV junction boxes (live in daylight).
- Do not disconnect d.c. plugs and sockets under load.

Commissioning test sheets must be completed showing which items have been inspected and which items on the sheet are not applicable to the particular installation (Fig. 1.50(a/b)).

On completion of the visual inspection the installation must be tested to ensure that it is safe and suitable for use. When carrying out testing on PV installations it is very important to remember that the d.c. side of the installation will be continually live and will present a risk of electric shock.

Documentation must be provided to show that both the d.c. and the a.c. parts of the microgeneration system have been satisfactorily tested.

PV COMMISSIONING TEST SHEET

Installation Check List

Installation Address	Inspected By	
	Date	Job number

Electrical

- ☐ Complies with standards
- ☐ Correctly selected an erected
- ☐ All equipment correctly connected
- ☐ Protective measures in place for special location if required
- ☐ Equipment suitable for external influences
- ☐ Installed to prevent mutual detrimental influence
- ☐ Conductors correctly identified
- ☐ Conductors correctly selected for current carrying capacity and volt drop
- ☐ All cables protected against mechanical damage or installed in safe zones
- ☐ Presence of fire barriers and protection against thermal effects

Mechanical

- ☐ Ventilation behind array to prevent overheating
- ☐ Array frame corrosion proof
- ☐ Array frame and material correctly fixed and roof weather proof
- ☐ Cable entry weatherproof

Protection against electric shock and overvoltage

- ☐ Basic protection provided for all live parts
- ☐ Protective bonding in place if required
- ☐ Surge protection provided if required
- ☐ RCD protection provided if required

d.c. part of the system

- ☐ Physical separation of d.c. and a.c. cables
- ☐ d.c. rated switch provided before inverter
- ☐ Correct cables used on d.c. side of the installation
- ☐ All d.c. components correctly rated for maximum system voltage (Voc stc x 1.25)
- ☐ PV string fuses and blocking diodes fitted if required

Figure 1.50 Commissioning sheet and array test.

PV array test

Job No	Installers name and address
Installation address	
Test date	Signature..
Description of work tested	Test instrument(s)

String		1	2	3	4	N
Array	Module					
	Quantity					
Array parameters	$V_{oc\ (stc)}$					
	$I_{sc\ (stc)}$					
Protective device	Type					
	Rating (A)					
	d.c. Rating (v)					
	Capacity (kA)					
Wiring	Type					
	Phase (mm^2)					
	Earth (mm^2)					
String test Test method	V_{oc} (V)					
	I_{sc} (A)					
	Sun					
Polarity check						
Earth continuity *If fitted*						
Connection to inverter *Serial number*						
Array insulation resistance test	Test Voltage (V)					
	Pos – Earth (MΩ)					
	Neg – Earth (MΩ)					
Comments on the installation						

Figure 1.50 *Continued*

Testing the d.c. side of the installation

The open circuit voltage (V_{oc}) of the installation must be measured and recorded. Unfortunately this is not simply a matter of measuring the voltage across the positive and negative cables at the d.c. isolator. Because the voltage of the panels are affected by temperature this also has to be taken into consideration, so this test is particularly important as it can be used to monitor whether the panels are becoming less efficient over a period of time (of course the first reading when they are new will be used as a reference point).

To carry out this test we must first look at the pv module datasheet to see what the temperature coefficient of voltage is for the module. It will be shown as temperature coefficient of V_{oc} (mV/°C).

All PV modules are tested under the same conditions, this enables us to make a true comparison between them when we are selecting the type which we are going to use. The temperature at which the panels are exposed to under standard test conditions (STC) is 25°C.

For use in this example let's say that the coefficient of V_{oc} is −104mV/°C. This figure means that each °C rise in temperature will result in a voltage loss of 104mV and each °C reduction in temperature will result in an increase in voltage of 104mV.

Now we must measure the temperature of the panel. If it is greater that the STC temperature which is 25°C then the expected voltage will be less. This is because the temperature rise will result in a loss of voltage, and of course if the measured temperature is less than STC then the expected voltage will be greater, as lower temperatures mean less voltage loss.

Measured module temperature for this example is 36°C and the V_{oc} which can be found on the module datasheet is 24 volts.

Step 1: Subtract the T_{stc} from the T_{actual}

$$36°C - 25°C = 11°C.$$

Step 2: Multiply the temperature difference by the temperature coefficient of V_{oc}

$$11°C \times \frac{104mV}{C} = 1144mV.$$

Step 3: Convert mV to volts by dividing by 1000

$$\frac{1144}{1000} = 1.144V.$$

Step 4: Subtract the volt drop from the module V_{oc}

$$24 - 1.144 = 22.855v.$$

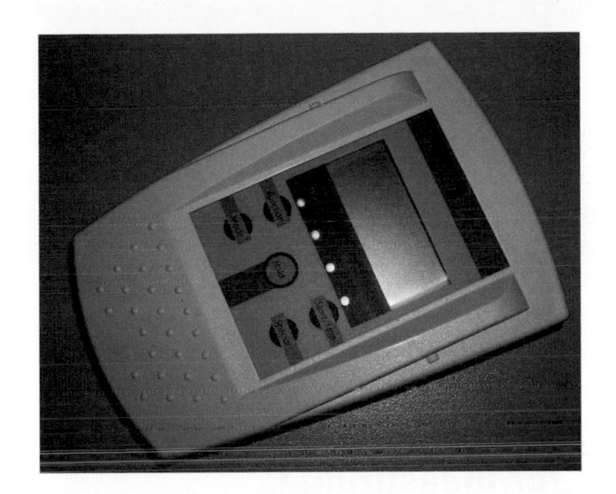

Figure 1.51 Irradiance metre.

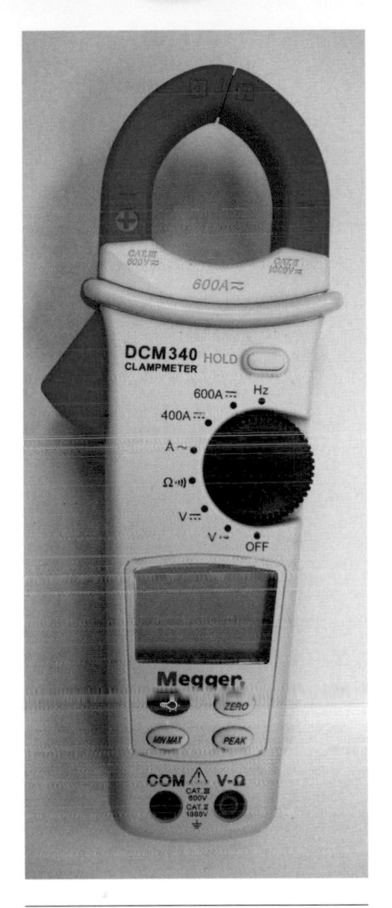

Figure 1.52 Clamp metre capable of measuring direct current.

This is the voltage which would be expected at each panel.

Step 5: Multiply the number of panels by the expected voltage. Lets say we have an array of 6 panels

$$6 \times 22.856\text{v} = 137.136\text{v}.$$

Step 6: Now we must measure the voltage at the d.c. isolator and it should be very close to our calculated value.

This type of calculation and measurement is important because if it is carried out regularly, say every two or three years, it will give an indication as to the efficiency of the system.

Another check which has to be carried out is a short circuit current test, this is to check that the array is producing the correct levels of current in relation to the level of light.

We must measure the irradiance levels using an irradiance metre (Fig. 1.51), and at the same time measure the short circuit current (I_{sc}) of the array using a DC clamp metre (Fig.1.52).

On a bright sunny day the measurements can be taken separately as the irradiance levels will be reasonably constant over a short period of time. However on a cloudy day getting an accurate measurement can prove difficult, and may involve two people. The irradiance value can change instantly and this of course will have an effect on the current being produced.

Having got both measurements the calculation is as follows:

Measured irradiance value = 80 W/m^2. Now divide this by 1000 to convert to kW = 0.080kW.

The standard test current for the array is 6 panels with a STC of 0.43A

$$6 \times 0.43 = 2.58.$$

Multiply the total STC by the measured irradiance in kW

$$2.58 \times 0.080 = 0.206\text{A}.$$

0.20A should be approximately the measured short circuit current.

All values used in the calculation are fictitious and have been used just to show how the calculation should be carried out. Actual measurements will vary from these depending on the type of modules, number of panels, etc.

Measuring the short circuit current of an array is quite a simple process but REMEMBER the d.c. side of the installation will be live during daylight hours.

Method for measuring I_{sc}

Isolate the a.c. side of the PV installation.

Isolate the d.c. side of the installation by using the d.c. switch disconnector nearest to the inverter.

Insert a shorting link across the positive and negative terminals on the dead side of the isolator, or join male and female connectors (Fig. 1.53).

Switch on the d.c. isolator and place a d.c. clamp metre around one of the incoming d.c. cables and record the reading.

Switch off d.c. isolator and remove links.

Switch the installation back on and leave in operating condition.

Protective earthing and bonding of the PV installation is normally not a requirement as the equipment used is usually class 2 equipment,

Figure 1.53 DC side of PV system shorted out and the current being measured by using a clamp metre.

however to reduce the risk of electric shock an insulation resistance test should be carried out.

Insulation resistance test

For this test an insulation resistance test instrument is used set at a voltage corresponding to the table (Fig. 1.54). This test is only to be carried out between live conductors and earth, NOT between live conductors.

Two methods can be used, one method is for the positive and negative of the PV installation to be connected together and a test must be carried out between these and any metal parts of the installation (such as the PV panel frame). A second test must be carried out between the joined conductors and the main installation earth. Of course where the panels are attached to a steel framed building which is bonded to the main earthing system it is only necessary to test between live conductors and the main installation earth.

It must always be remembered that the panels are working during daylight and that the pv conductors will be live. The test procedure would be

- Turn off the d.c. isolator.
- Disconnect the PV cables from the inverter.
- Join together the positive and negative conductors on the isolated side of the system.
- Connect the one lead of the insulation resistance test instrument to the joined PV conductors and connect the other lead to earth.
- Switch on the d.c. isolator and carry out the test, I would always test at 250 volts first and then if the reading was satisfactory I would then test at 500 volts.
- After completing the test, turn off the d.c. isolator.

Test method	System voltage (Voc stc x 1.25)	Test voltage	Minimum resistance value
Array positive and negative shorted together	120V	250V	0.25MΩ
	<600 V	500V	0.5MΩ
	<1000V	1000V	1MΩ
Test between positive and negative separately	120V	250V	0.25MΩ
	<600	500V	0.5MΩ
	1000V	1000V	1MΩ

Figure 1.54 Insulation resistance test voltage and minimum resistance values.

- Disconnect test leads.
- Reconnect PV cables and inverter.

The other method is to carry out the test using the positive and negative separately.

Where the test is being carried out as described in the second method, precautions must be taken to ensure that the test voltages do not exceed the module or cable rating.

Testing the a.c. side of the installation

For all new connections to a supply system, an initial verification must be carried out on the a.c. part of the PV installation. On completion an Electrical Installation Certificate (Fig. 1.55) must be completed, along with a schedule of test results (Fig. 1.56) and a schedule of inspections (Fig. 1.57). These documents must only be completed by a person who is registered to self certificate electrical work which has been carried out by them.

In installations where it has been chosen to connect the PV circuit by using a spare way in an existing consumer unit, it is a requirement that the earthing arrangements for the existing installation are upgraded to comply with the latest edition of BS 7671 wiring regulations.

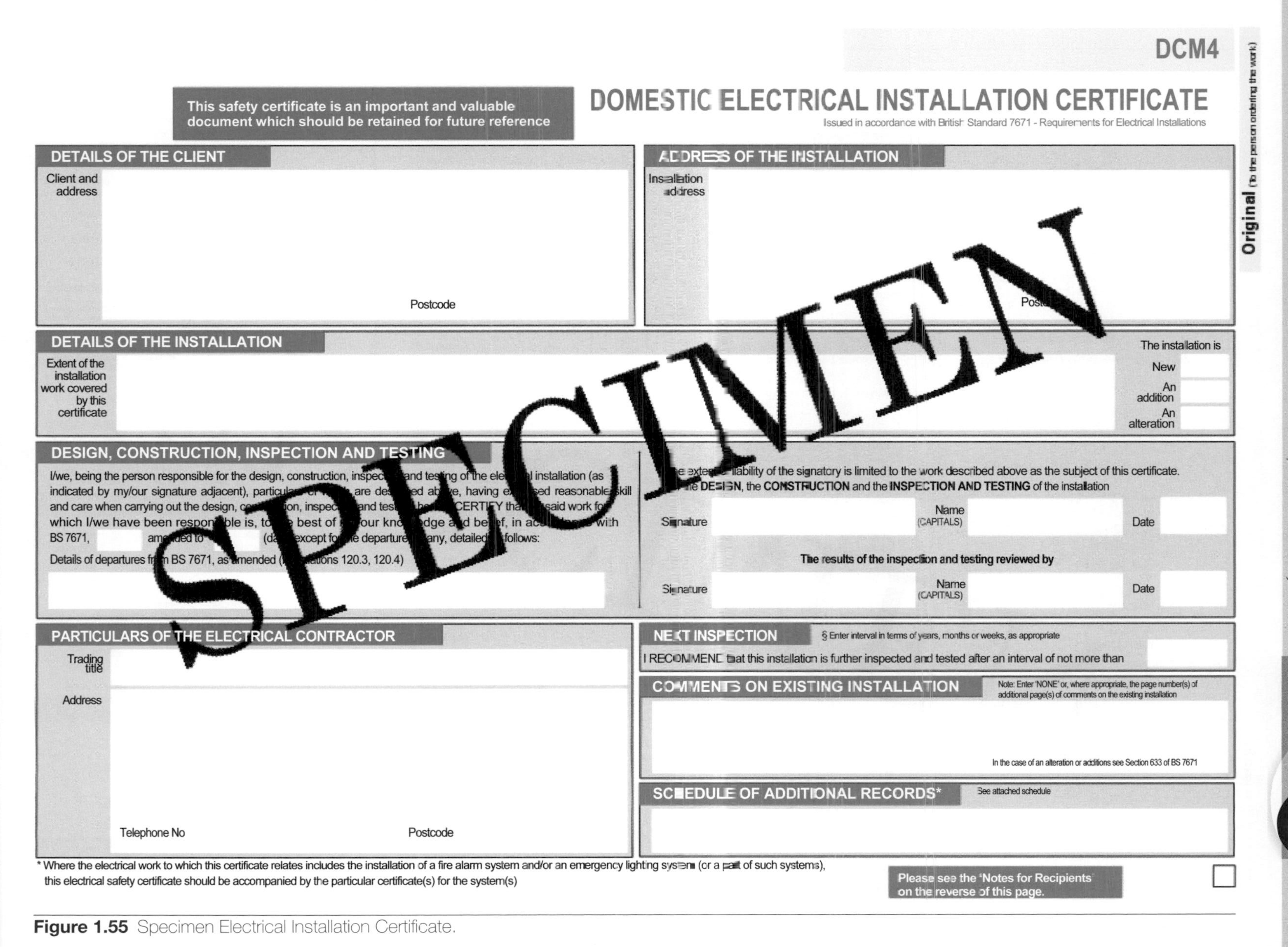

DCM4

Original (To the person ordering the work)

This safety certificate is an important and valuable document which should be retained for future reference

DOMESTIC ELECTRICAL INSTALLATION CERTIFICATE

Issued in accordance with British Standard 7671 - Requirements for Electrical Installations

DETAILS OF THE CLIENT

Client and address

Postcode

ADDRESS OF THE INSTALLATION

Installation address

Post[illegible]

DETAILS OF THE INSTALLATION

Extent of the installation work covered by this certificate

The installation is

New

An addition

An alteration

DESIGN, CONSTRUCTION, INSPECTION AND TESTING

I/we, being the person responsible for the design, construction, inspec[illegible] and testing of the ele[illegible] installation (as indicated by my/our signature adjacent), particula[illegible] are des[illegible] ab[illegible]e, having ex[illegible] reasonable skill and care when carrying out the design, c[illegible]on, inspec[illegible] and tes[illegible] CERTIFY tha[illegible] said work for which I/we have been respon[illegible]ble is, to [illegible] best of [illegible] our kno[illegible]dge a[illegible]d be[illegible]f, in ac[illegible] with BS 7671, ______ amended to ______ (da[illegible] except fo[illegible] departure[illegible] any, detailed [illegible] follows:

Details of departures from BS 7671, as amended ([illegible] 120.3, 120.4)

[illegible] extent [illegible] liability of the signatory is limited to the work described above as the subject of this certificate.

[illegible] the **DESIGN**, the **CONSTRUCTION** and the **INSPECTION AND TESTING** of the installation

Signature | Name (CAPITALS) | Date

The results of the inspection and testing reviewed by

Signature | Name (CAPITALS) | Date

PARTICULARS OF THE ELECTRICAL CONTRACTOR

Trading title

Address

Telephone No

Postcode

NEXT INSPECTION § Enter interval in terms of years, months or weeks, as appropriate

I RECOMMEND that this installation is further inspected and tested after an interval of not more than

COMMENTS ON EXISTING INSTALLATION Note: Enter 'NONE' or, where appropriate, the page number(s) of additional page(s) of comments on the existing installation

In the case of an alteration or additions see Section 633 of BS 7671

SCHEDULE OF ADDITIONAL RECORDS* See attached schedule

* Where the electrical work to which this certificate relates includes the installation of a fire alarm system and/or an emergency lighting system (or a part of such systems), this electrical safety certificate should be accompanied by the particular certificate(s) for the system(s)

Please see the 'Notes for Recipients' on the reverse of this page.

Figure 1.55 Specimen Electrical Installation Certificate.

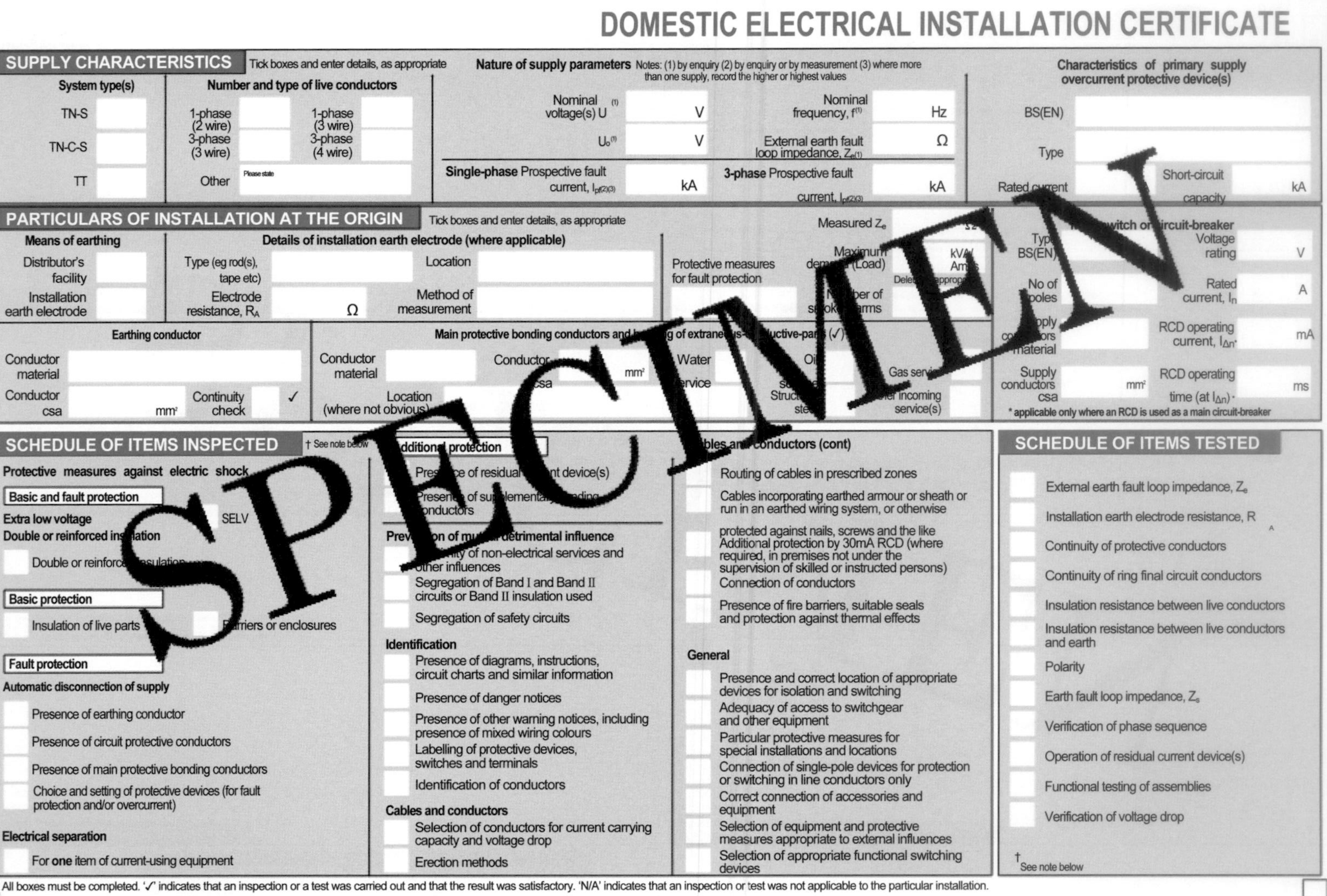

DCM4

Original (To the person ordering the work)

DOMESTIC ELECTRICAL INSTALLATION CERTIFICATE

SUPPLY CHARACTERISTICS

Tick boxes and enter details, as appropriate

System type(s): TN-S / TN-C-S / TT

Number and type of live conductors: 1-phase (2 wire) / 1-phase (3 wire) / 3-phase (3 wire) / 3-phase (4 wire) / Other (Please state)

Nature of supply parameters Notes: (1) by enquiry (2) by enquiry or by measurement (3) where more than one supply, record the higher or highest values

Nominal voltage(s) U (1) ... V
U_0 (1) ... V
Nominal frequency, f (1) ... Hz
External earth fault loop impedance, Z_e (1) ... Ω
Single-phase Prospective fault current, I_{pf} (2)(3) ... kA
3-phase Prospective fault current, I_{pf} (2)(3) ... kA

Characteristics of primary supply overcurrent protective device(s)

BS(EN)
Type
Rated current
Short-circuit capacity ... kA

PARTICULARS OF INSTALLATION AT THE ORIGIN

Tick boxes and enter details, as appropriate

Means of earthing: Distributor's facility / Installation earth electrode

Details of installation earth electrode (where applicable)

Type (eg rod(s), tape etc)
Location
Electrode resistance, R_A ... Ω
Method of measurement

Measured Z_e
Maximum demand (Load) ... kVA/Amps (Delete as appropriate)
Protective measures for fault protection
Number of smoke alarms

...witch or circuit-breaker
Type BS(EN)
Voltage rating ... V
No of poles
Rated current, I_n ... A
Supply conductors material
RCD operating current, $I_{\Delta n}$* ... mA
Supply conductors csa ... mm²
RCD operating time (at $I_{\Delta n}$)* ... ms
* applicable only where an RCD is used as a main circuit-breaker

Earthing conductor

Conductor material
Conductor csa ... mm²
Continuity check ✓

Main protective bonding conductors and bonding of extraneous-conductive-parts (✓)

Conductor material
Conductor csa ... mm²
Location (where not obvious)
Water service
Oil service
Gas service
Structural steel
Other incoming service(s)

SCHEDULE OF ITEMS INSPECTED

† See note below

Protective measures against electric shock

Basic and fault protection
Extra low voltage
SELV
Double or reinforced insulation
Double or reinforced insulation

Basic protection
Insulation of live parts
Barriers or enclosures

Fault protection
Automatic disconnection of supply
Presence of earthing conductor
Presence of circuit protective conductors
Presence of main protective bonding conductors
Choice and setting of protective devices (for fault protection and/or overcurrent)

Electrical separation
For **one** item of current-using equipment

Additional protection
Presence of residual current device(s)
Presence of supplementary bonding conductors

Prevention of mutual detrimental influence
Proximity of non-electrical services and other influences
Segregation of Band I and Band II circuits or Band II insulation used
Segregation of safety circuits

Identification
Presence of diagrams, instructions, circuit charts and similar information
Presence of danger notices
Presence of other warning notices, including presence of mixed wiring colours
Labelling of protective devices, switches and terminals
Identification of conductors

Cables and conductors
Selection of conductors for current carrying capacity and voltage drop
Erection methods

Cables and conductors (cont)
Routing of cables in prescribed zones
Cables incorporating earthed armour or sheath or run in an earthed wiring system, or otherwise protected against nails, screws and the like
Additional protection by 30mA RCD (where required, in premises not under the supervision of skilled or instructed persons)
Connection of conductors
Presence of fire barriers, suitable seals and protection against thermal effects

General
Presence and correct location of appropriate devices for isolation and switching
Adequacy of access to switchgear and other equipment
Particular protective measures for special installations and locations
Connection of single-pole devices for protection or switching in line conductors only
Correct connection of accessories and equipment
Selection of equipment and protective measures appropriate to external influences
Selection of appropriate functional switching devices

SCHEDULE OF ITEMS TESTED

External earth fault loop impedance, Z_e
Installation earth electrode resistance, R_A
Continuity of protective conductors
Continuity of ring final circuit conductors
Insulation resistance between live conductors
Insulation resistance between live conductors and earth
Polarity
Earth fault loop impedance, Z_s
Verification of phase sequence
Operation of residual current device(s)
Functional testing of assemblies
Verification of voltage drop

† See note below

† All boxes must be completed. '✓' indicates that an inspection or a test was carried out and that the result was satisfactory. 'N/A' indicates that an inspection or test was not applicable to the particular installation.

‡ This form is based on the model Electrical Installation Certificate shown in Appendix 6 of BS 7671(as amended).
Where a smoke alarm has been installed, separate certification is required on the appropriate form.

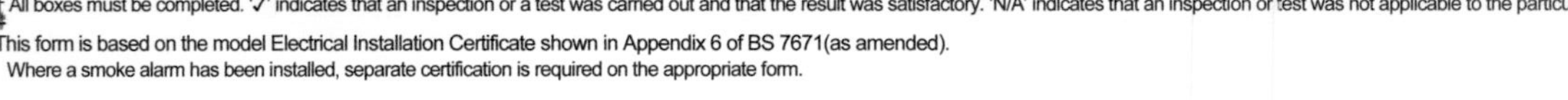

Figure 1.56 Specimen Electrical Installation Certificate: schedule of test results.

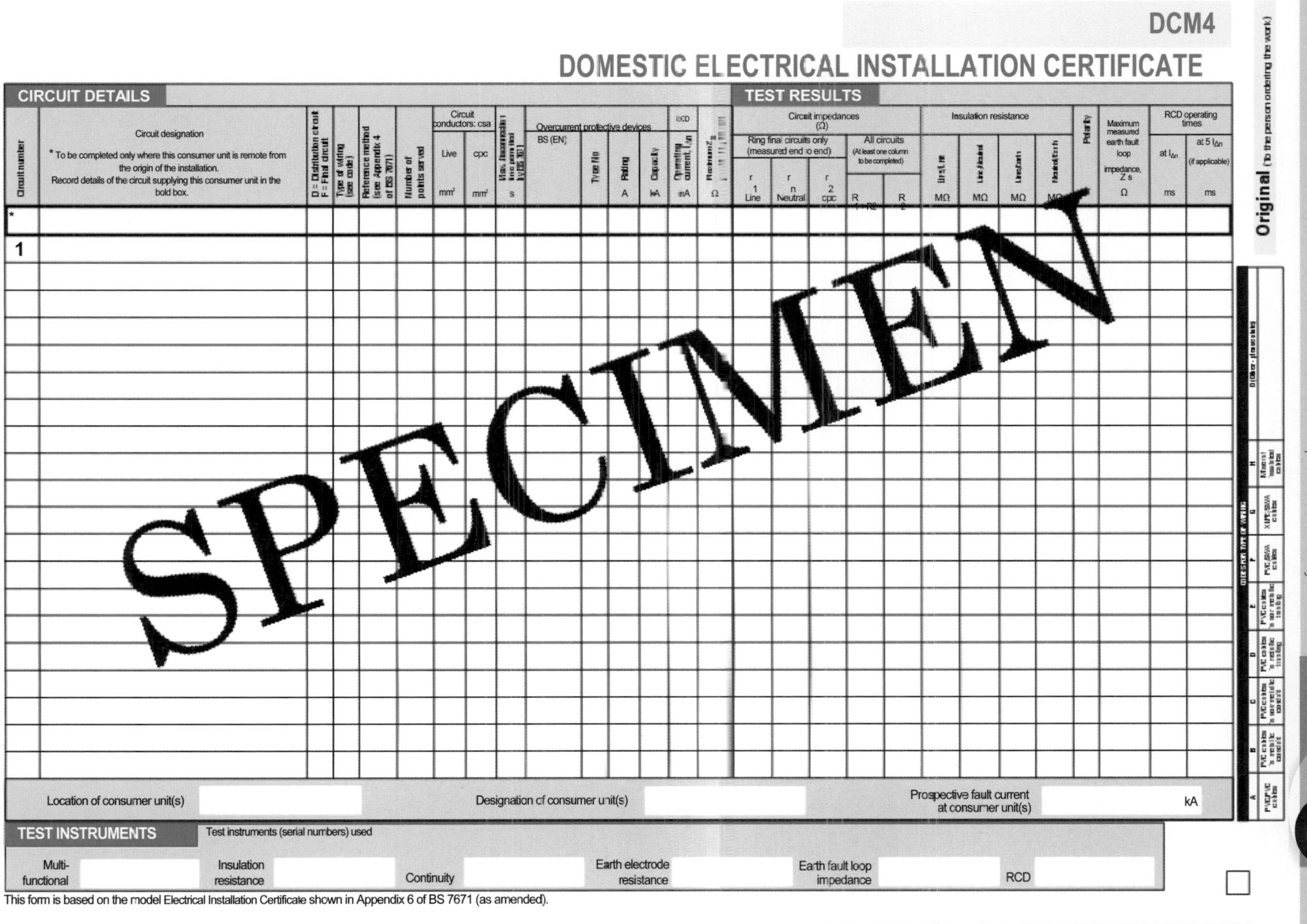

DCM4

DOMESTIC ELECTRICAL INSTALLATION CERTIFICATE

Original (To the person ordering the work)

CIRCUIT DETAILS															TEST RESULTS												
						Circuit conductors: csa			Overcurrent protective devices				RCD		Circuit impedances (Ω)					Insulation resistance						RCD operating times	
															Ring final circuits only (measured end to end)			All circuits (At least one column to be completed)									
Circuit number	Circuit designation * To be completed only where this consumer unit is remote from the origin of the installation. Record details of the circuit supplying this consumer unit in the bold box.	D = Distribution circuit F = Final circuit	Type of wiring (see code)	Reference method (see Appendix 4 of BS 7671)	Number of points served	Live mm²	cpc mm²	Max. disconnection time permitted by BS 7671 s	BS (EN)	Type No	Rating A	Capacity kA	Operating current, $I_{\Delta n}$ mA	Maximum Z_s Ω	r_1 Line	r_n Neutral	r_2 cpc	R_1+R_2	R_2	Line/Line MΩ	Line/Neutral MΩ	Line/Earth MΩ	Neutral/Earth MΩ	Polarity	Maximum measured earth fault loop impedance, Z_s Ω	at $I_{\Delta n}$ ms	at 5 $I_{\Delta n}$ (if applicable) ms
*																											
1																											

SPECIMEN

Location of consumer unit(s)		Designation of consumer unit(s)		Prospective fault current at consumer unit(s)	kA

TEST INSTRUMENTS Test instruments (serial numbers) used

Multi-functional	Insulation resistance	Continuity	Earth electrode resistance	Earth fault loop impedance	RCD

CODES FOR TYPE OF WIRING

A	B	C	D	E	F	G	H	O
PVC/PVC cables	PVC cables in metallic conduit	PVC cables in non-metallic conduit	PVC cables in metallic trunking	PVC cables in non-metallic trunking	PVC/SWA cables	XLPE/SWA cables	Mineral insulated cables	Other - please state

This form is based on the model Electrical Installation Certificate shown in Appendix 6 of BS 7671 (as amended).

Figure 1.57 Specimen Electrical Installation Certificate: schedule of inspection.

CHAPTER 2

Solar thermal heating

Solar thermal systems use the energy from the sun to heat water which can then be stored and used as required.

In countries which have a hot climate, and even on nice sunny days in the UK, the simplest way to provide hot water from the sun is to drape a hosepipe over your roof (Fig. 2.1), or even across a patio.

It is surprising how quickly the water heats up and how hot it gets. Of course this method works but it is not really very practical as it is very uncontrollable, and in this basic form it is not possible to store the hot water.

If we were to place a coil of the hose into a vessel containing water the hose would work as a heat exchanger, just as a coil of copper does inside a hot water cylinder. Figure 2.2 shows a cut away cylinder with the coil exposed.

In this way we could store the water and use it when we needed it. Provided that the vessel was well insulated to prevent the heat from the stored water escaping, it could be stored for quite some time. This of course is the basic principle of solar thermal hot water systems. Over many years it has evolved to become quite an efficient and clean way to provide hot water to domestic and commercial buildings.

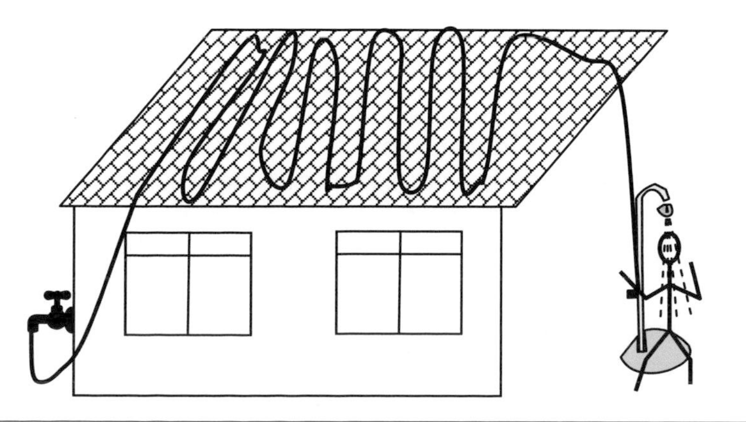

Figure 2.1 Simple method of using the solar energy provided by the sun.

Figure 2.2 A typical hot water cylinder with a primary coil used as a heat exchanger.

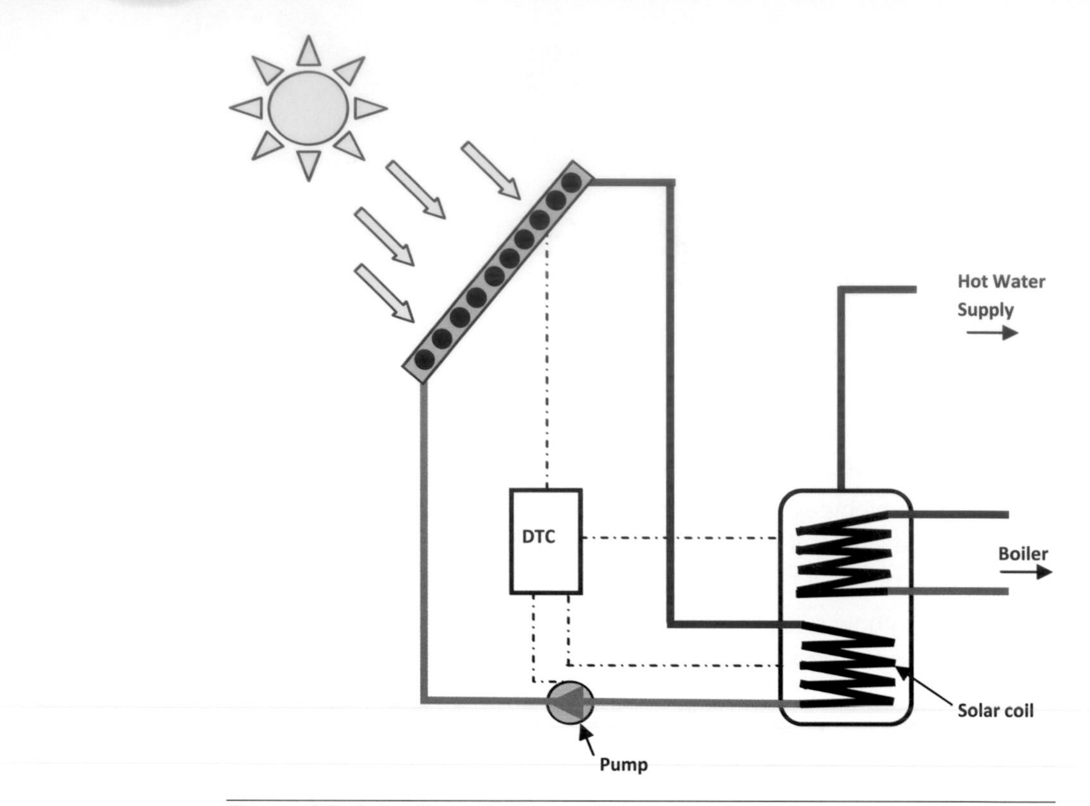

Figure 2.3 Active thermal system using a twin coil cylinder.

The part of the solar thermal system which collects the heat from the sun is known as the primary side of the system and the water which is heated by the solar system is known as the secondary side. This is the domestic hot water side of the installation.

In modern systems a solar thermal heat collector is used to collect heat from the sun and then feed it into a cylinder via a heat exchanger. The heat collector is usually fitted onto a roof, although it can be simply installed at ground level in a garden if required.

As with anything which relies on heat or light from the sun, it is far better if the collector is facing directly at it. As a solar collector is usually fixed in one position it is important to make sure that it is the best possible position, facing south at an angle of 35° would be the ideal position as it will ensure the highest average of sunlight is directed onto the collector. Figure 2.4a is a drawing showing the azimuth angle of the sun and Figure 2.4b shows the horizontal angle of the sun.

Of course this is the best position but not every roof faces south at an angle of 35° from horizontal. For this reason a site survey should always be carried out just to ensure that the siting of the solar collector will allow it to provide enough energy to make the installation worthwhile.

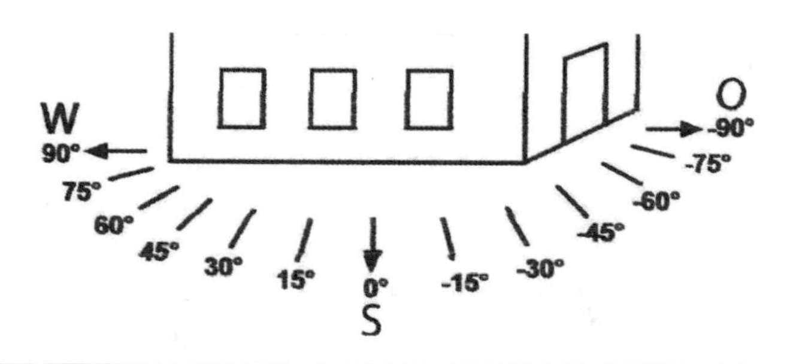

Figure 2.4 (a) As the sun moves across the sky the angle between it and the solar panel alters, if the panel faces south it will see the sun all day.

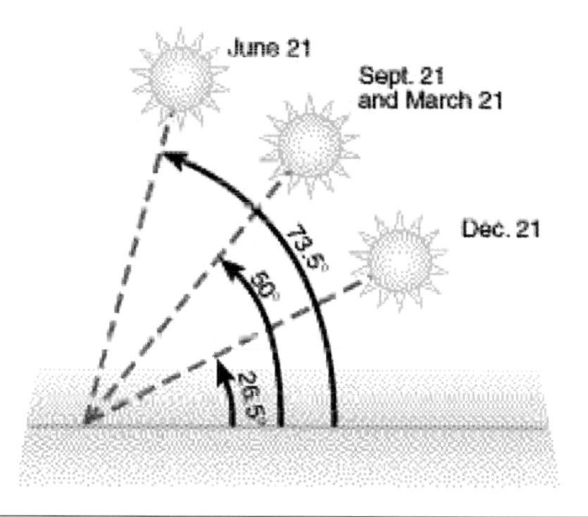

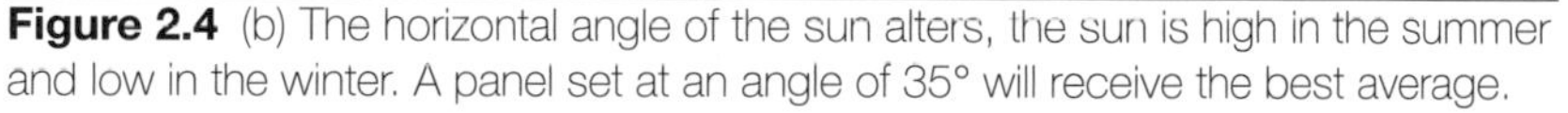

Figure 2.4 (b) The horizontal angle of the sun alters, the sun is high in the summer and low in the winter. A panel set at an angle of 35° will receive the best average.

It is worth remembering that a collector which is sited facing 35° east or west of due south will have a system loss of around 10% efficiency. The angle from horizontal is not quite so critical and often a steeper angle may be beneficial, particularly in urban areas where a lot of dust is present, as at a steep angle the rain will wash the dust away and help to keep the surface of the collector clean.

On dwellings which do not face south there are other options which could be considered. For instance, the collector area could be increased to gain a better output when the sun is shining on to it. Or two or more collectors could be used, one facing east for the morning sun and another facing west for the afternoon and evening sun; this of course depends on the orientation of the roof.

Types of collectors

The most common types of solar thermal collector used in domestic installations are either *flat plate* or *evacuated tube collectors*.

Flat plate collector

A flat plate collector (Fig. 2.5a) consists of a plate which is usually copper, on top of which a tubing system is fixed. All of this is then painted black and placed into an insulated box with a glass cover. This all sounds very simple and in its basic form it is, but of course it would not be as efficient as we would like it to be.

Modern technology allows us to improve the efficiency by using special materials. The bottom and the sides of the unit containing the collector would be very well insulated, using insulating materials capable of withstanding the high temperatures which could be as high as 300°C on a very hot day.

The paint used to coat the flat plate and the tubing would be non-reflective to ensure that as much heat as possible is kept within the collector, it would also need to be able to withstand the high temperatures.

Special glass and glass coatings have also been developed which in simple terms allows the radiation from the sun to pass through it but does not allow it back out again; this type of glass is said to have a selective surface. It is not uncommon for the glass cover to be double glazed as this also reduces the heat lost from inside the unit; the downside of this is that a double glazed unit would also reduce the incoming light. Low iron glass is also used as it allows for good light transmission.

Connection to the solar thermal panel should be made using materials which are suitable to withstand the very high temperatures which will be present. Figure 2.5b is a stainless steel flexible pipe which would be suitable for use in solar thermal systems.

Soldered fittings must not be used as they would become unstable during periods of very high temperatures.

Once fitted into place the tubing is filled with a fluid which will not freeze in cold conditions; remember the collector is usually on a roof

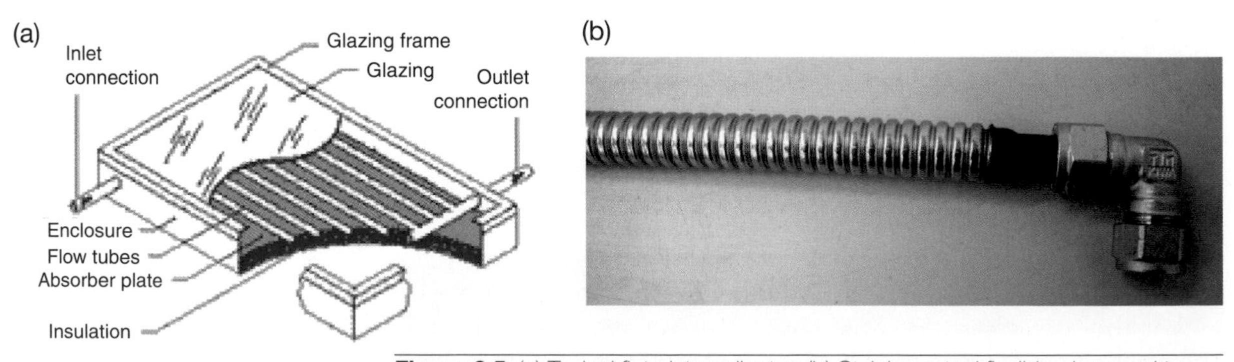

Figure 2.5 (a) Typical flat plate collector. (b) Stainless steel flexible pipe used to connect the solar panel.

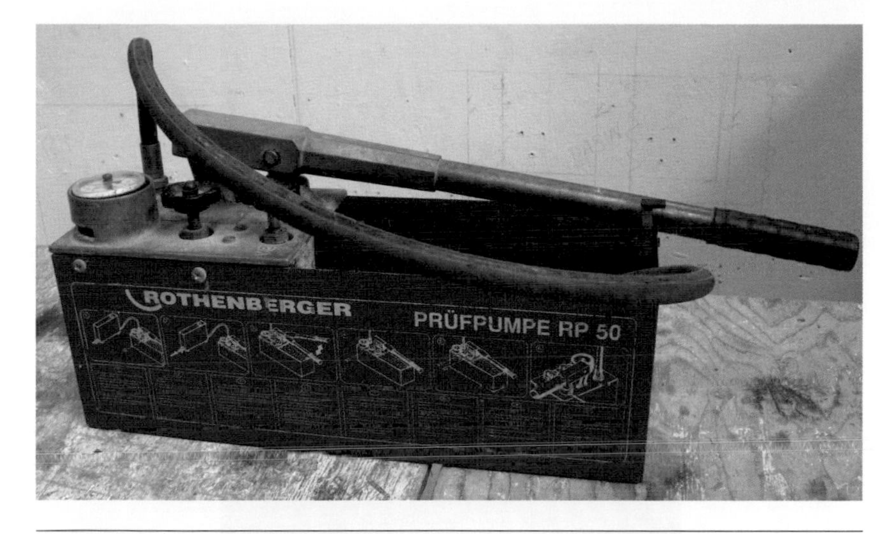

Figure 2.6 A pressure test pump.
Source: Sovello

which will get very cold in the winter and if the system was to freeze all sorts of problems would arise. For instance when the fluid freezes it expands, this could rupture the tubing or at the very least block the tubing and prevent circulation; if expansion routes were to become blocked there is a strong possibility of explosion. For this reason the fluid must contain antifreeze. The type of antifreeze used by motorists is not suitable for solar thermal installations as it is very toxic and must not be used anywhere that it could get into our domestic water system; the solution will also be subjected to high temperatures.

The most common type of solution is aqueous polypropylene glycol. This is a high temperature, low toxicity chemical which is ideally suited for use in solar thermal systems. It is also used as a heat transfer fluid in heat pump collectors. As with any type of heating system the use of corrosion inhibitors is also important, as antifreeze used by itself will increase the risk of corrosion and it is important that the system is designed and installed to work for many years without needing major repairs. Before filling any system it must be pressure tested to ensure that there are no leaks. A pressure test can be carried out by using a pressure test pump (Fig. 2.6) or by using the solar filling/flushing pump.

Evacuated tube collector

There are two different types of evacuated tube collector, a heat pipe collector and a non-heat pipe collector. Both of them require the use of a number of glass tubes which have to be made of very tough glass such as pyrex, and each tube has a vacuum created inside it and this is why it is known as an evacuated tube.

Figure 2.7 An evacuated tube solar thermal panel.
Source: Shutterstock

The vacuum inside the tube prevents losses from convection, and this is the same principle used to keep food or fluids hot in a vacuum flask. The tube usually has a selective coating which allows the radiation to pass through it but stops it reflecting back out again; this is the same type of coating which a flat plate collector would have.

Heat pipe system

A heat pipe collector has a metal tube, usually copper which is filled with a small amount of liquid such as purified water with some special additives, or more commonly alcohol, as this vaporises easily at low temperature. This tube has a large fin connected to it which acts as a heat exchanger. The metal tube with the fin connected is inserted into the glass tube, the tube is then evacuated and sealed with the metal tube protruding from one end.

When the tube is exposed to solar radiation (sunlight) the fin heats up which in turn heats the metal tube. The heat causes the liquid in the tube to vaporise and rise to the exposed metal tip of the tube, which then gets very hot. Figure 2.8a shows the exposed metal tip which fits into the manifold and Figure 2.8b shows the cross section of an evacuated tube.

Figure 2.8 (a) The tip of the heat pipe gets very hot

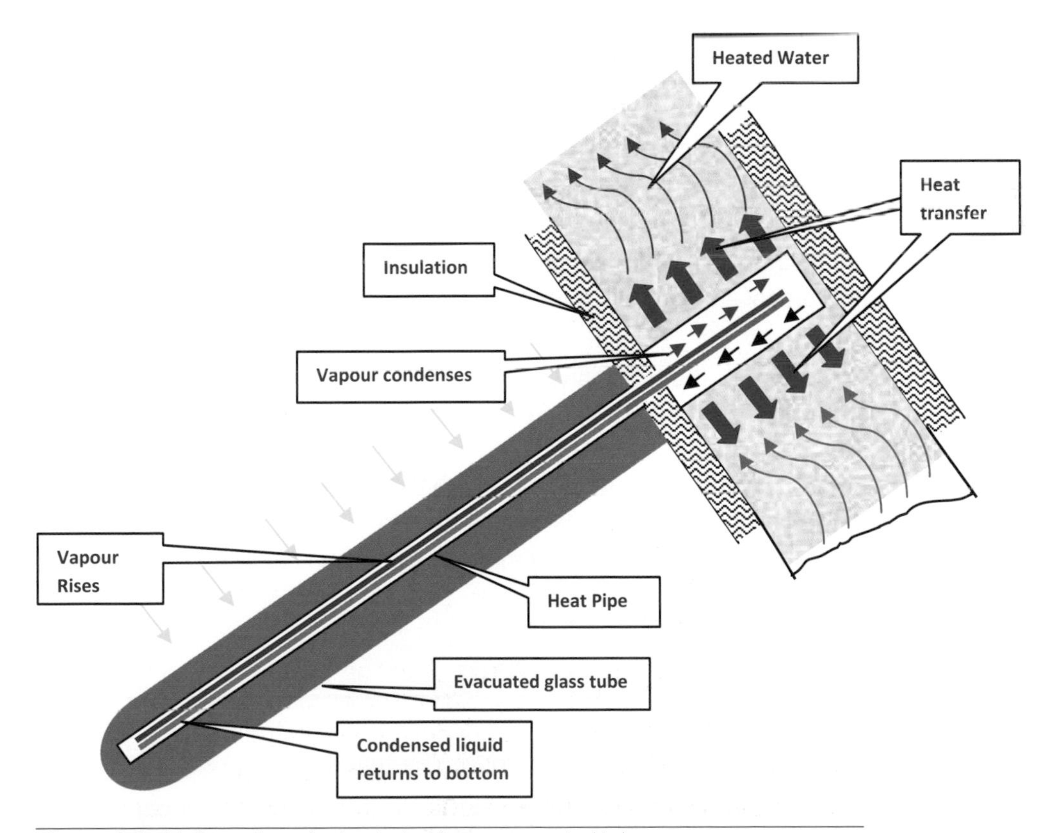

Figure 2.8 (b) Operation of a heat pipe in an evacuated tube.

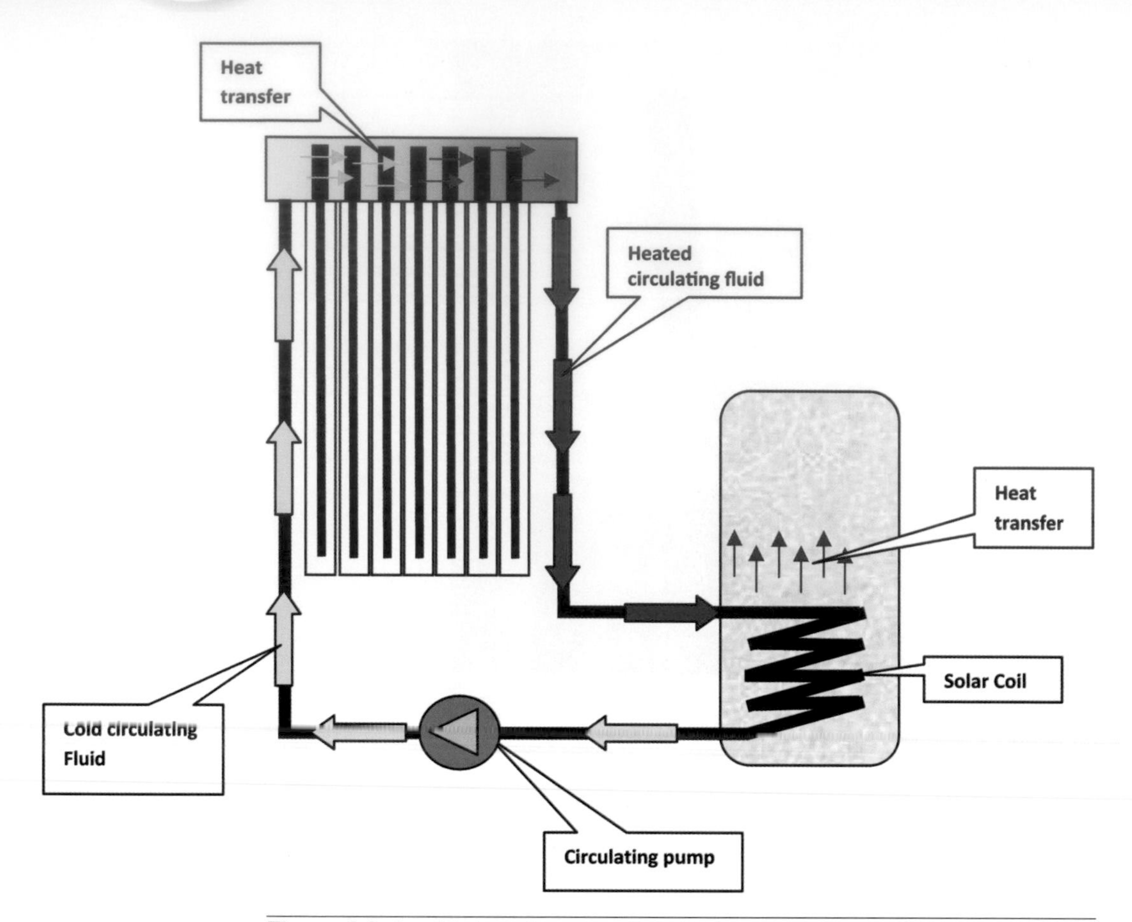

Figure 2.9 A collection of evacuated tubes forming a solar thermal panel which heats up the circulating fluid in a solar system.

It is important that the heat pipe evacuated tubes are installed as near to vertical as possible, if they are laid flat the vapour will not be able to rise.

A number of the tubes are then used to make a solar thermal collector: the protruding end of the metal tubes are inserted into a manifold, the circulating fluid used in the primary side of the solar thermal system then passes through the manifold, over the hot ends of the metal tubes and heats up (Fig. 2.9).

The other type of evacuated tube solar thermal collector is a non-heat pipe system.

Non-heat pipe system

This type of system allows the circulating fluid in the primary side of the heating system to pass through the tubes.

The evacuated tube looks very similar to that used in the heat pipe system, the difference being that the steel tube is not a sealed tube

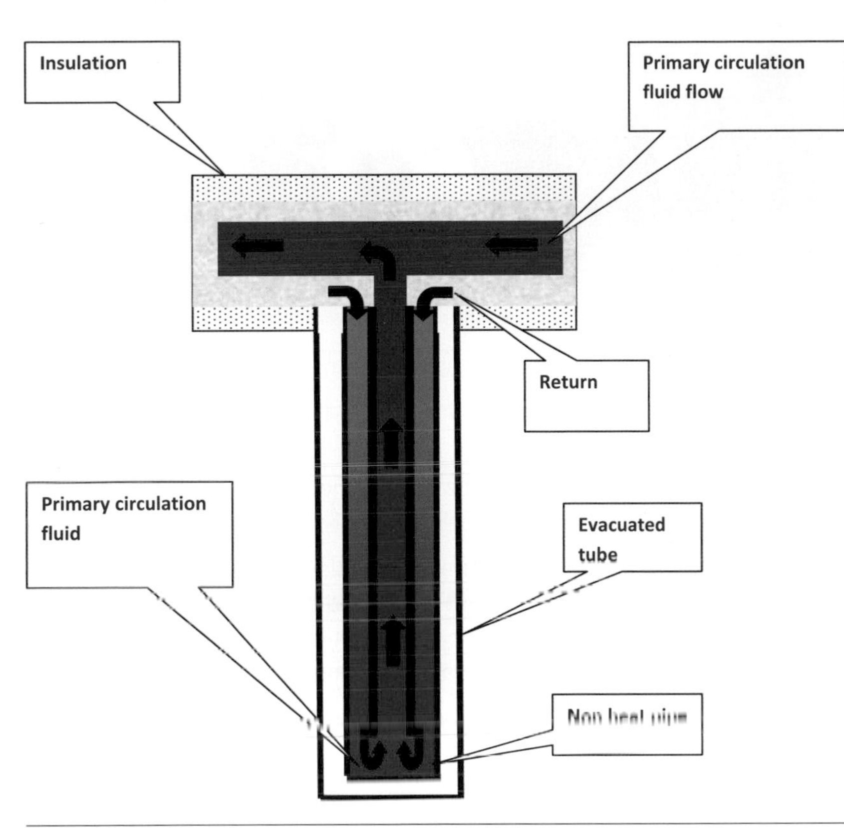

Figure 2.10 A non-heat pipe in an evacuated tube.

with liquid in it but a tube within a tube which allows the primary circulating fluid to flow through it (Fig. 2.10).

This of course means that the primary fluid is heated directly from the sun's radiation rather than the heat being transferred to it via a heat exchanger. In this type of collector the manifold as well as the collector tube has to be a tube within a tube, the outer section of the manifold is the return and the inner section is the flow. As can be seen from the drawing, the circulating fluid would flow through the outer tube of the manifold into the outer section of the tube which is in the evacuated tube. The heat which is collected from the sun's radiation by the fins attached to the tube would then be absorbed by the circulating fluid, the fluid would then flow through the inner section of the tube back into the inner section of the manifold. Due to the amount of heat which the circulating fluid will be subjected to on hot days it is important that the correct type of fluid is used; protection from freezing is also vital in any type of solar thermal system.

Propylene glycol is ideal for solar thermal installations as it is food safe. If by a system failure any of the fluid should leak into the domestic water system it would not result in a major health hazard,

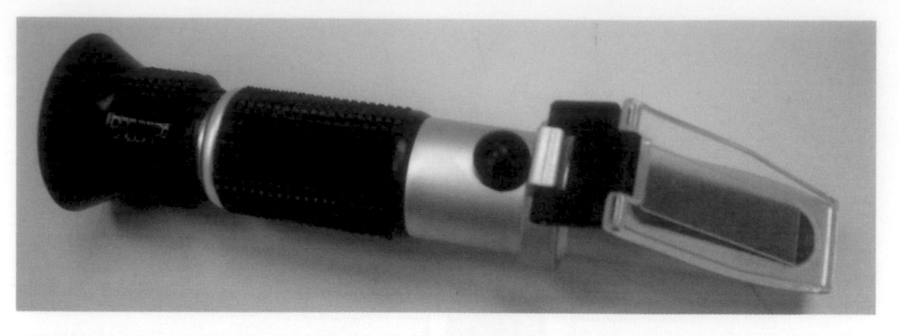

Figure 2.11 It is important to ensure that the system has the correct ratio of water to glycol.

unlike the problems which would arise if other types of antifreeze were to be used.

The glycol should be diluted with water at just slightly less than 50/50, and the level of glycol must be tested on a regular basis, usually annually. The test can be carried out by using an instrument called a refractometer (Fig. 2.11): a small amount of fluid is placed onto the glass plate at the front of the tool, it is then viewed though the eyepiece and an indication of the ratio of water to glycol will be seen.

Types of solar thermal systems

All-electric buildings

Although solar thermal systems produce a great deal of free energy which is used to heat our water, it is very unlikely that a solar thermal system installed in the UK would provide enough hot water to fulfil the requirements for a domestic installation each and every day, particularly during the winter months. For this reason solar thermal systems need a backup system which can be used to supplement the domestic hot water requirements.

A cylinder with a coil for the solar primary circuit and an immersion heater installed which operated when required would be a very simple method and is often used where the building is all-electric.

In this type of system a control is used which is called a differential temperature controller (DTC). This device recognises when the temperature in the solar collector is less than that in the bottom of the cylinder. In this situation if the circulating pump was to keep running it would have the effect of removing the heat from the cylinder and using it to heat the collector. This would very quickly cool down the hot water cylinder, leaving a cylinder full of cold water.

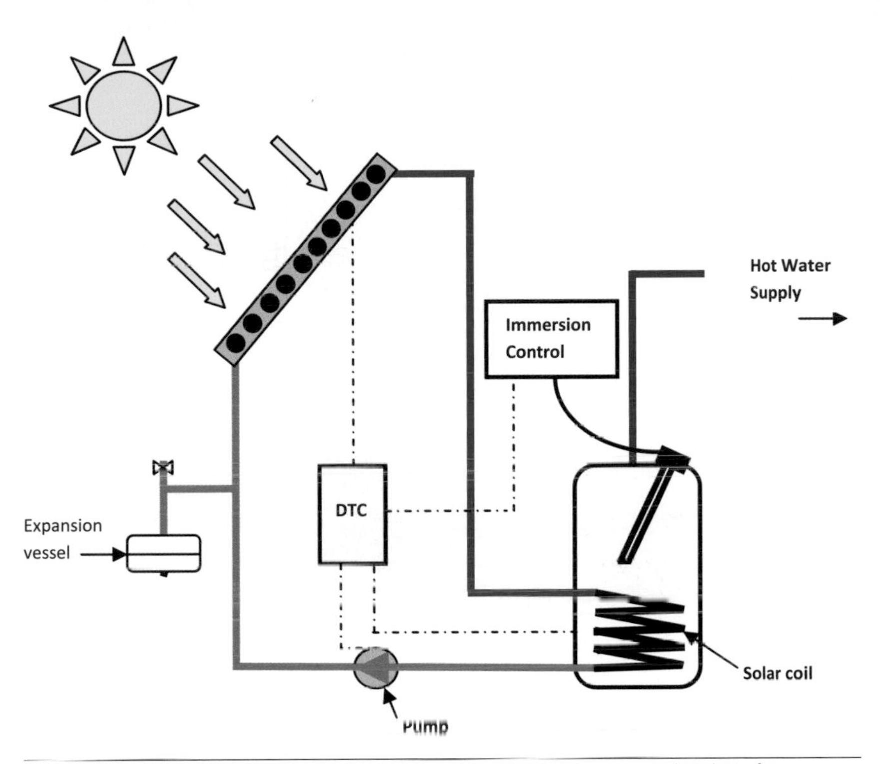

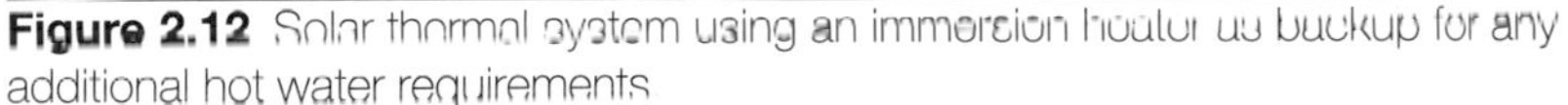
Figure 2.12 Solar thermal system using an immersion heater as backup for any additional hot water requirements

When the DTC recognises that the water in the collector is lower than that in the primary coil, in the bottom of the cylinder, the circulating pump is automatically switched off and the immersion heater is switched on, unless of course the water in the cylinder is at the required temperature. The DTC will also be set to ensure that the temperature of the domestic hot water reaches 60°C for at least 1 hour in every 24-hour period to eliminate the risk of legionnaires disease. Figure 2.12 shows the basic principle of how the system will work, but it must be remembered that when water heats it will expand and provision for this expansion must be provided by using an expansion vessel.

Buildings with other heating provision

In systems which use solar thermal for the domestic hot water and a boiler fired by fossil fuels for the central heating it is usual to have a cylinder which utilises two coils. One coil (the lower one) is used for the solar thermal circuit and a second coil is used by the boiler system to supplement the hot water requirement when required. See Figure 2.13.

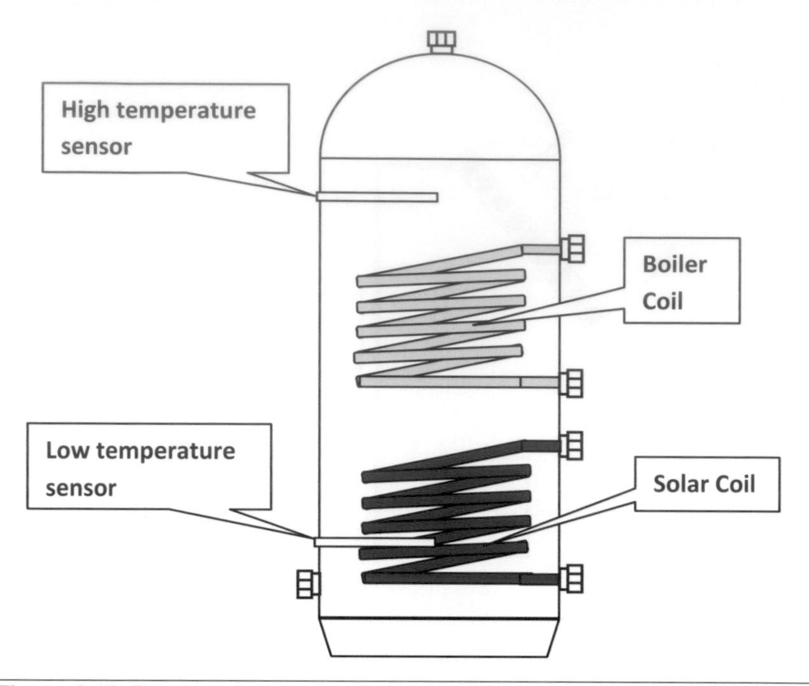

Figure 2.13 Drawing of a twin coil cylinder, the bottom coil would be used for the solar thermal part of the system.

Although all types of solar thermal systems work using the same basic principle, there are several different installation methods for these systems, the use of which depends on the site conditions.

Drainback system

The simplest of these systems is a known as a drainback system. In this type of system sensors are installed which sense the temperature of the solar collector and the temperature of the water in the bottom of the DHW cylinder. When the sensor in the collector detects useful thermal heat the circulating pump will start and pump the circulating fluid through the collector. As the circulating fluid passes through the collector the heat will be transferred from the collector tubes into the fluid. This fluid will then pass through a drainback vessel and then into the primary coil (heat exchanger) which is in the bottom of the hot water cylinder where the heat will be transferred to the water in the cylinder. The circulation will continue until the sensors detect that the water in the cylinder is hotter than the water in the solar collector. Figure 2.14 shows the drainback system in operation when the collector is hotter than the cylinder.

At this point the pump will stop and the circulating fluid will drain back into the drainback vessel situated inside the building. In some cases this will eliminate the need for antifreeze solution to be used

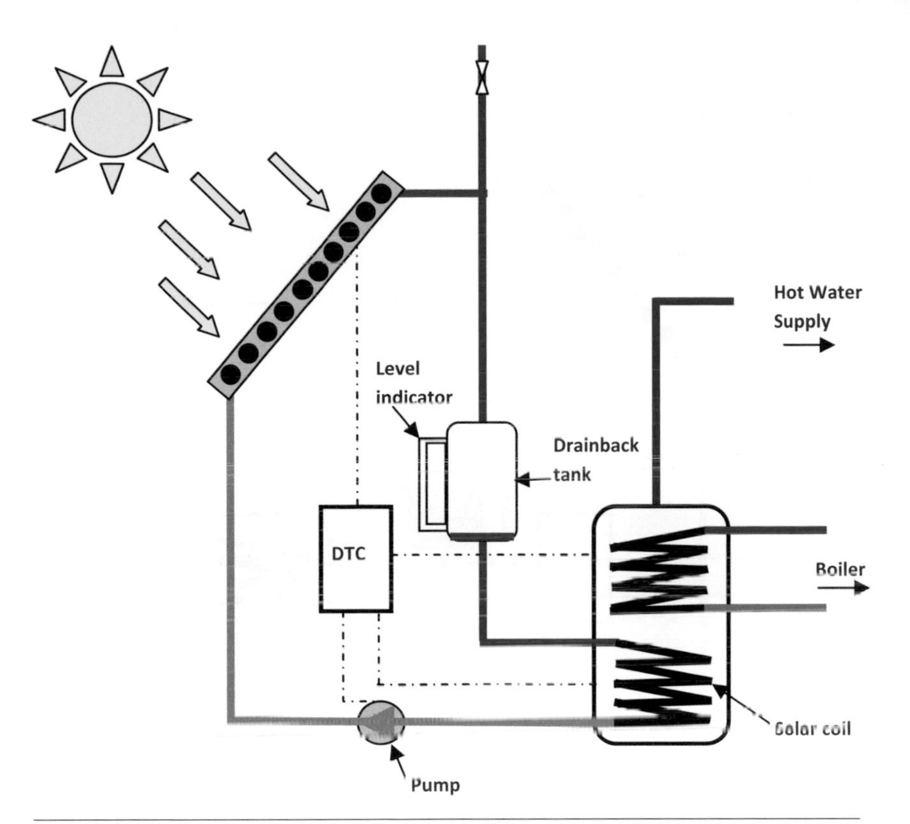

Figure 2.14 Drawing of a typical drainback system in operation.

as the circulating fluid, although it is generally advisable that the fluid does have additives added to it as apart from preventing the fluid from freezing, it has other properties which help the system with regards to scale prevention and boiling. Figure 2.15a shows the drainback system at standstill when the cylinder is hotter than the collector.

As can be seen from the drawing, this system is sealed, and because of this the vessel which is used for the fluid to drain back into will also need to be large enough to act as an expansion vessel which will allow the volume of fluid to expand and contract as the temperature of the fluid changes. Figure 2.15b is of a small drainback vessel.

There are a couple of important points to remember with this type of system. One is that the drainback vessel needs to be positioned higher than the pump, this is to ensure that it is never left dry. If a pump starts with no water in it on a regular basis the life expectancy of the pump will be greatly reduced. Secondly, the type of collector used must be suitable as when the hot water requirement for the cylinder is satisfied and the collector is empty it will get very hot. The collector manufacturer must be consulted as to its suitability.

(a)

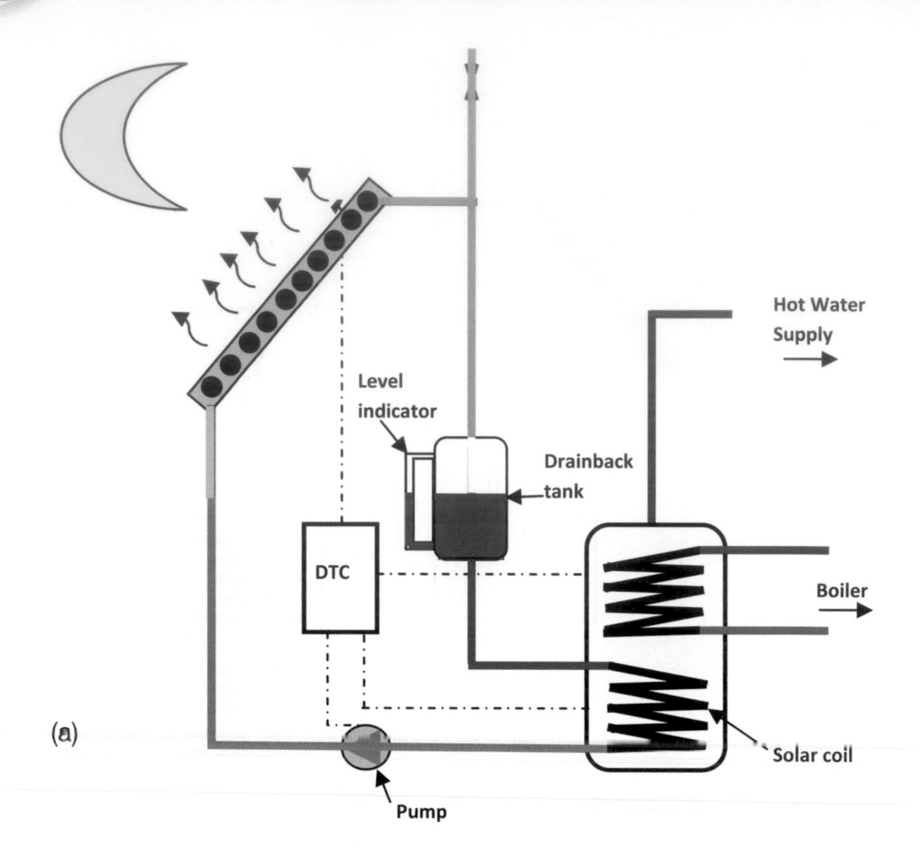

(b)

Figure 2.15 (a) Drainback system where there is no solar radiation on the solar panel. The circulating fluid has drained back into the drainback vessel. (b) A small drainback vessel.

Open vented indirect system

This system works as a conventional non-sealed system where the collector remains full of the heat transfer fluid at all times. Any expansion in the system fluid is via an expansion pipe into a header tank, this header tank also ensures that the system is topped up as required. A twin coiled cylinder is used with the lower coil being used

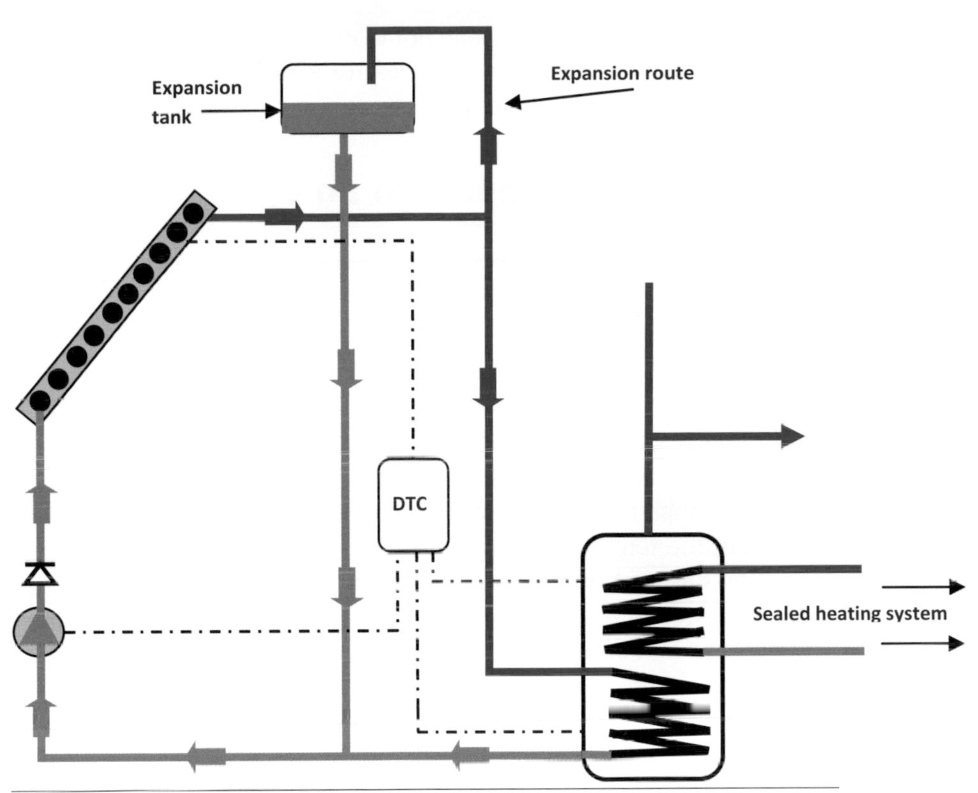

Figure 2.16 Open vented system.

for the solar thermal part of the system and the top coil being used for the boiler part of the installation (Fig. 2.16).

When the hot water requirement of this type of system is satisfied the pump will stop working and the heat transfer fluid will simply sit in the system, this condition is known as stagnation.

Clearly if the fluid is not moving and the sun is shining onto the collector the fluid will get very hot. If the correct type of fluid is used, i.e. polypropylene glycol, the boiling point of the fluid will rise, which will reduce the chance of the fluid boiling, although the boiling point in an open vented system will be lower than it would be in a pressurised system.

It will not be possible to prevent the fluid from boiling entirely. The problem that arises when the fluid boils is that the glycol, over a period of time will degrade and become less effective as an antifreeze. For this reason it is important that annual checks are made, usually in the autumn, to ensure that the correct level of glycol in the system is maintained to prevent the system from freezing in the winter. Where this type of system is installed it can operate alongside either an open vented or an unvented sealed heating system. The principle of the solar system is the same as the other systems, they all have a differential temperature controller (DTC) which detects when the domestic hot water is calling for heat, and when the temperature in the solar collector is higher than that of the water in the bottom of the storage cylinder.

Sealed solar system with an open vented cylinder

In this system the solar thermal part of the installation is sealed and pressurised and the cylinder is open vented, the heating part of the system can be either vented or unvented.

A sealed solar thermal system works very well as it is operating under pressure, usually around 1 bar which is equivalent to 14.69 lbs per square inch (psi). As the system is sealed it must be fitted with an expansion vessel which will allow the fluid to expand and contract at the temperature changes. Figure 2.17a is of a typical expansion vessel, but of course they come in all shapes and sizes. Many vessels use a diaphragm and others may use a balloon. Figure 2.17b shows a cut away section of a balloon type. Great care must be taken to ensure that any vessel used is the correct size for the system.

Figure 2.18 is a simple drawing of how a diaphragm type expansion vessel would work. Vessel A is where the system is not under pressure and the air in the vessel is not compressed and vessel B shows the system under pressure. When used on solar thermal

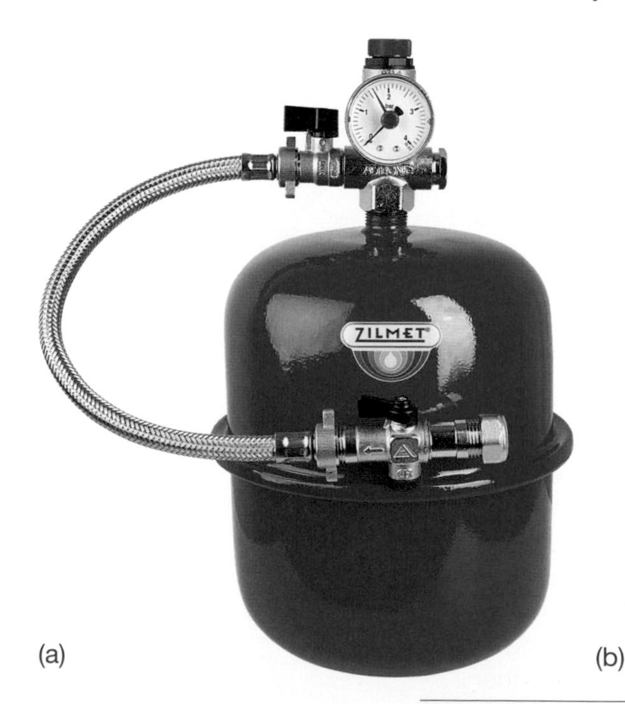

(a)

(b)

Figure 2.17 (a) Typical expansion vessel. (b) Cut away section of balloon of a type.

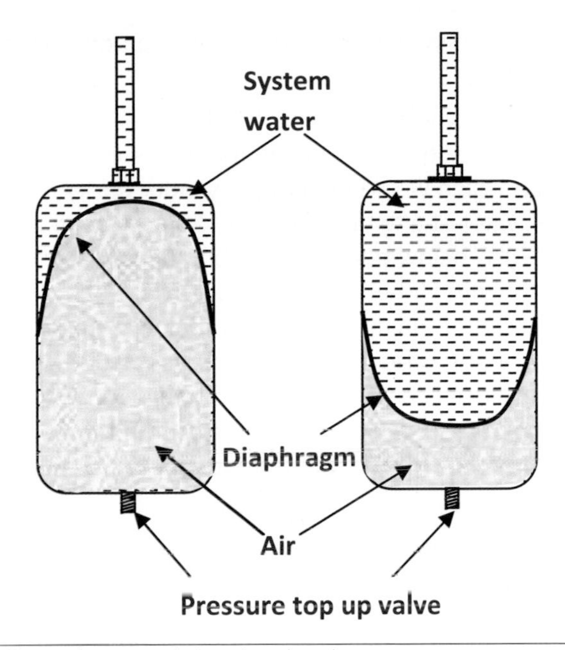

Figure 2.18 Expansion vessel at rest and under pressure.

systems it is usual to fit the pressure vessel with the water inlet at the top, the reason for this is to help prevent the diaphragm from being damaged by very hot water or even steam in some cases.

All pressurised systems must be fitted with a pressure gauge and a pressure relief valve.

Because the system is sealed there is no loss of fluid due to evaporation, although as with a sealed central heating system a facility must be provided for topping the system up if required. A differential temperature controller is used in this system just as it is in the other systems. Figure 2.12 is of a sealed solar system.

Where a sealed system is used a pipe must be run from the pressure relief valve into a container which is large enough to hold the entire volume of antifreeze solution which is in the primary part of the solar system, this is to catch the fluid in the event of the relief valve operating.

Unvented sealed systems

An unvented completely sealed system is probably the most common and most efficient system which could be installed, not only is the solar part of the system sealed and pressurised, the heating and domestic hot water part of the system is as well.

With all sealed systems we must always remember that the boiling point of the fluid under pressure is higher than when it is not under pressure, the table (Fig. 2.19) shows the boiling points of water at different pressures.

Clearly any system which is under pressure must be suitably designed with all of the correct safety measures in place. Imagine the devastation which would occur if a cylinder under a pressure of 3 bar with the water at a temperature of 120°C were to rupture. This is nearly 20°C below boiling point, but if this water is suddenly released into the atmospheric pressure outside of the cylinder, or the pressure inside of the cylinder falls to below 2 bar, it will boil instantly. When water boils its volume increases 1600 times and it would be just like a bomb going off. Not only would the clothes in the airing cupboard get wet, they would end up three streets away along with most of the materials which the house was made of.

Because of the dangers involved with unvented heating systems all installers must be suitably qualified in the installation and

Pressure		Boiling Point	
psi	bar	deg F	deg C
0.5	0.03	79.6	26.4
1	0.07	102	38.7
2	0.14	126	52.2
3	0.21	141	60.8
4	0.28	153	67.2
5	0.34	162	72.3
6	0.41	170	76.7
7	0.48	177	80.4
8	0.55	183	83.8
9	0.62	188	86.8
10	0.69	193	89.6
11	0.76	198	92.1
12	0.83	202	94.4
13	0.90	206	96.6
14	0.97	210	98.7
14.69	**1.0**	**212**	**100**
15	1.0	213	101
16	1.1	216	102
17	1.2	219	104
18	1.2	222	106
19	1.3	225	107
20	1.4	228	109
22	1.5	233	112
24	1.7	238	114
26	1.8	242	117
28	1.9	246	119
30	2.1	250	121
32	2.2	254	123
34	2.3	258	125
36	2.5	261	127
38	2.6	264	129
40	2.8	267	131
42	2.9	270	132
44	3.0	273	134
46	3.2	276	135
48	3.3	279	137

Figure 2.19 Table of boiling points in relation to pressure.

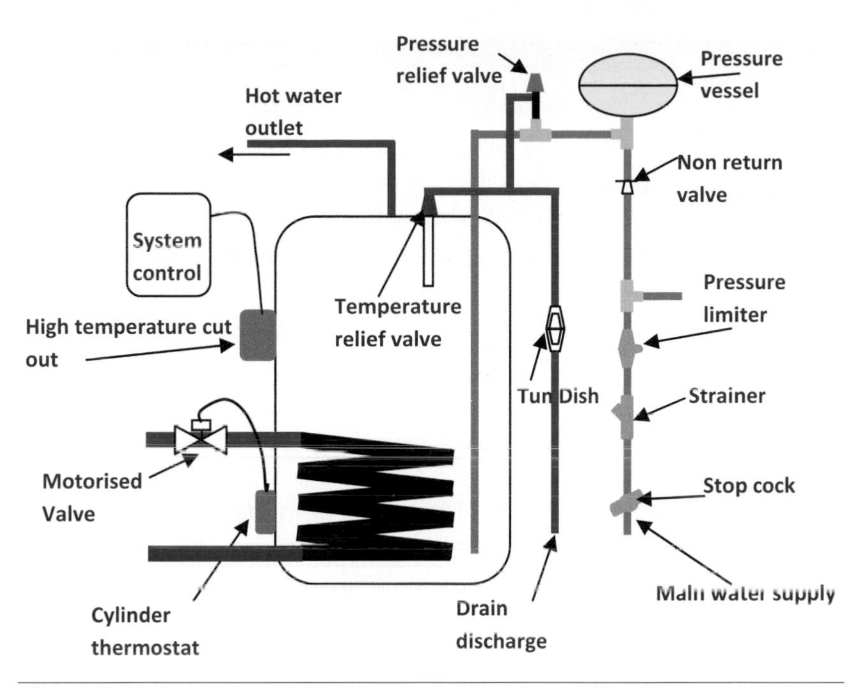

Figure 2.20 Protection requirements for an unvented heating system

commissioning of them. The systems must have also have safety devices fitted which give three tiers of protection (Fig. 2.20).

The first tier is a device called a cylinder thermostat which cuts off the heat source to the cylinder when it has reached its required temperature. This is usually set around 60–65°C and it also switches the heat source back on again when the cylinder temperature drops.

The second tier is a device called a high temperature energy cut out, and is set at between 80°C and 85°C. The function of this device is to shut down the heating system completely if the cylinder thermostat fails. The cut out must not be self resetting, it must only be able to be reset manually. This of course it to prevent any malfunction of the system going unnoticed.

The third tier is a pressure relief valve which would be set slightly above the operating pressure of the system. If the first two levels of protection were to fail the pressure relief valve would open and allow the water to discharge, this in turn would relieve the pressure in the system and prevent it from exploding. An important part of this pressure relief system is known as a tundish, which provides an air gap between the discharge pipe from the cylinder or system and the pipe which discharges the fluid to the outside of the building. An air gap has a dual function: it allows a person to see that fluid is being discharged and it also provides a route for the fluid if the pipework outside becomes blocked or frozen.

Filling the system

With any sealed system the filling process has to be carried out carefully. It is very important that all of the air is removed from the system and if the system is a pressurised one, the pressure has to be correct.

If the system is fitted with an expansion vessel, the expansion vessel must be charged to the system operating pressure before filling the system up. This is done with a bicycle pump and a pressure gauge (Fig. 2.21a/b).

Once the expansion vessel has been charged, the easiest method of flushing and filling the system is to use an electric solar filling pump. This piece of equipment will allow the flushing and filling to be carried out without difficulty. The tubing from the filling pump must be connected to the fill points of the solar control (Fig. 2.22).

Another method would be to use a hand filling pump (Fig. 2.23). This is a perfectly satisfactory method although it will take a little longer than using an electric filling pump.

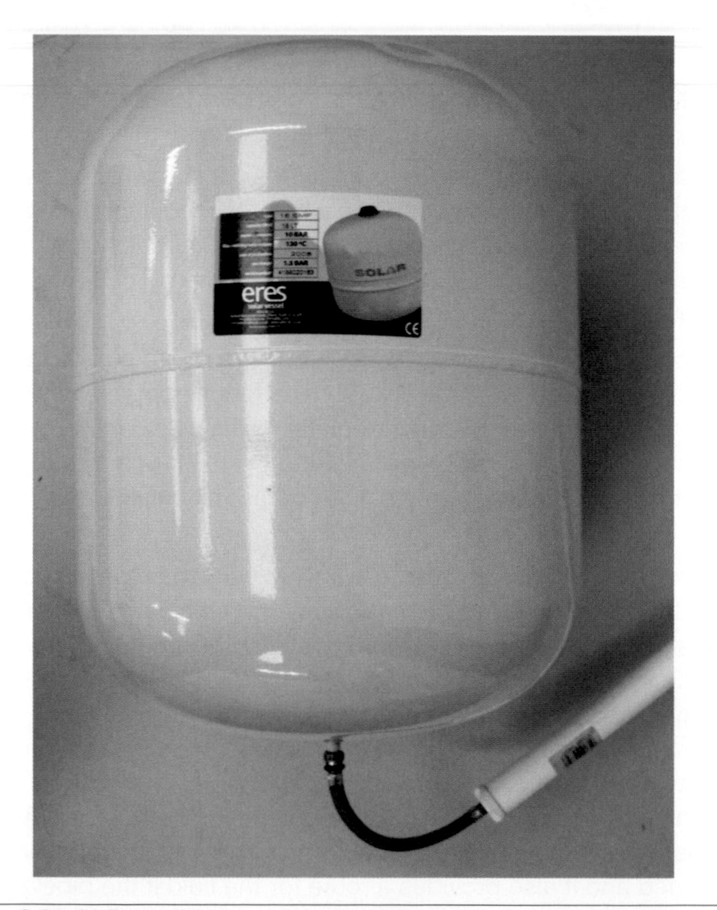

Figure 2.21 (a) Pressurising an expansion vessel.

Figure 2.21 (b) Checking an expansion vessel for the correct pressure.

Figure 2.22 Flexible pipes used to fill the system connected to the solar pumping station.

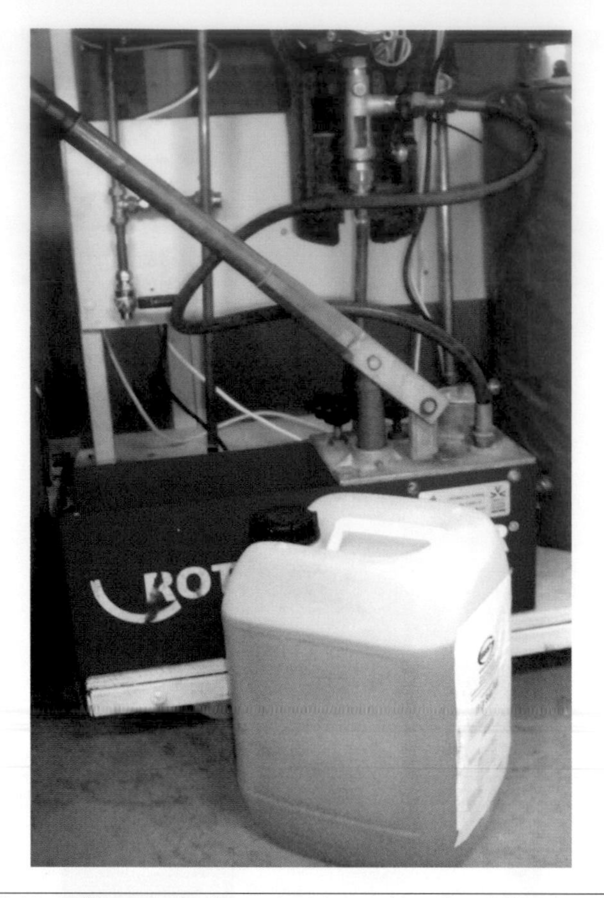

Figure 2.23 Hand filling a solar thermal system using a hand filling/pressure pump.

As previously mentioned the system must be designed to allow for flushing, filling and topping up. All that is required is that a tube can be connected to each end of the system to enable the fluid to flow into one end, right around the system and back out of the other end. The fitting of a filling valve, which are made specifically for filling these types of system, is often the easiest option. Most solar control stations will have the required filling points (Fig. 2.22) which will allow the fluid to be directed around the system as required.

Before filling the system a pressure test must be carried out. The easiest method is to fill the system with water using the fill pump (Fig. 2.23). Figures 2.24 (a) and (b) show the valve positions for filling the system.

Once the system is filled the filling valve on the return side of the system must now be turned off and the fill pump allowed to keep running until the test pressure is reached (Fig. 2.25).

This should be slightly over the system operating pressure, usually a test pressure of 3/4 bar would be suitable although this does depend on the size of the system. It is always best to consult the manufacturer's

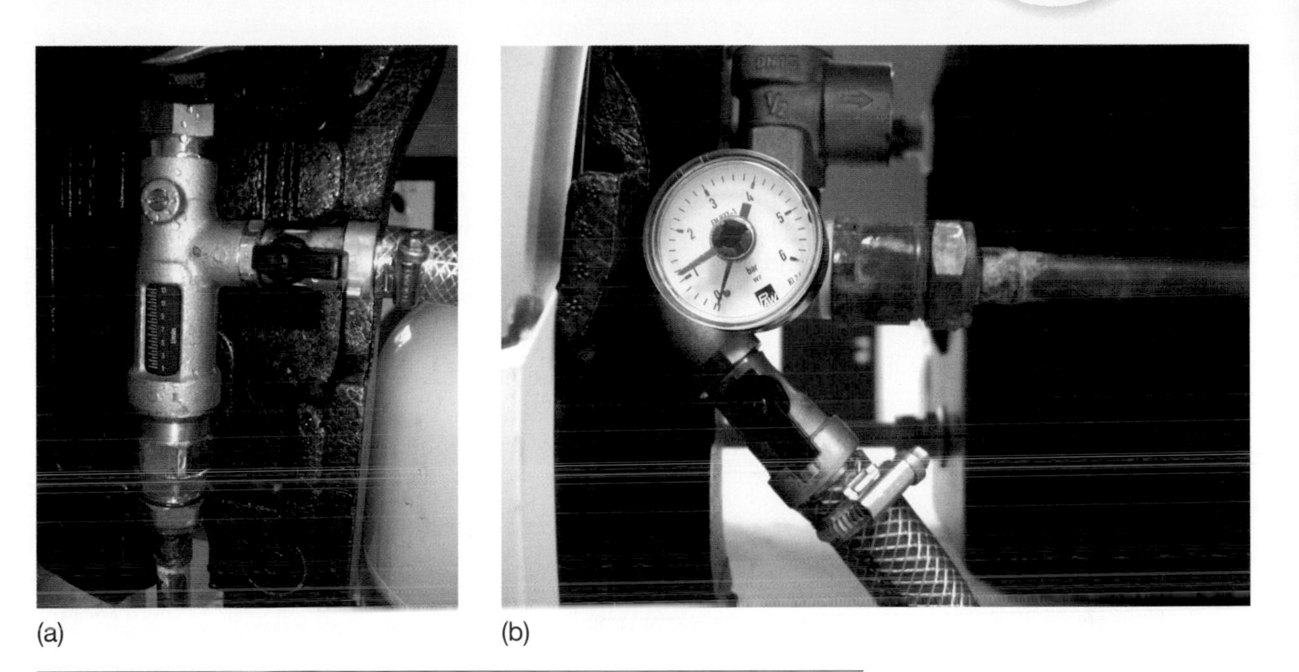
(a) (b)

Figure 2.24 Valve positions for filling the system.

instructions. Once at the required pressure the flow side of the filling valve should be closed and the filling pump turned off, the system should now be left for an hour or so to see if the pressure drops.

Once the system has been pressure tested it must be flushed thoroughly to remove any dirt/debris which may have been introduced

Figure 2.25 Return valve turned off/system filled until the correct operating pressure is reached.

into the system during the installation process. Flushing will also remove any air pockets which may be in the system. The fluid used for flushing is a mixture of water and a chemical specifically developed for this purpose. This chemical can also be used in the pressure testing process as it will make life easier as the filling pump reservoir can be filled with the solution at the beginning of the process. Both flow and return ports must be opened on the fill valve and the bypass part of the valve must be closed as shown in Fig. 2.24a/b.

The filling pump must then be allowed to run for an hour or so to ensure that all of the dirt and air is flushed out of the system. One important thing to remember is that the fill pump reservoir must not be allowed to run dry as this will allow air to be sucked into the system.

When the flushing process is complete the flow and return ports of the fill valve must be closed and the pump turned off. The fill pump reservoir must be emptied of the flushing fluid and filled with the glycol mix which is going to be used in the solar thermal system. Disconnect the return pipe from the reservoir and place the end into an empty vessel or drain (it is important to consult the manufacturer's instructions to make sure that the fluid is suitable to be discharged directly into a drain). The pump must now be turned on and both flow and return ports of the fill valve opened, the fill pump reservoir must be kept topped up with the glycol mixture, and the end of the return pipe must be monitored to see when the glycol mixture has circulated right through the system. Once the glycol has circulated, the return pipe must be placed into the fill pump reservoir and the pump left to circulate the fluid for a minute or two. Then any air vents on the system must be opened to bleed any remaining air out of the system.

The system must now be pressurised: the return port of the fill valve must be closed and the bypass port opened (Fig. 2.26).

Figure 2.26 Return port closed with bypass port opened.

Figure 2.27 Correct pressure reached with fill valve closed.

The pump will now increase the pressure to the required level, but as soon as the system is at the correct pressure the flow port of the fill valve must be closed (Fig. 2.27) and the fill pump turned off.

Any surplus glycol which is in the fill pump reservoir can now be placed into a clean container and put aside for future use, such as topping up the system. It is not unusual for the pressure to drop in a new system after a few days due to air in the system fluid.

CHAPTER 3

Domestic wind turbines

For domestic use wind turbines come in two categories, microgeneration and small scale generation. Microgeneration will produce up to 1.5kW of energy while small-scale generation will produce from 1.5kW to 15kW.

Although here in the United Kingdom we get 40% of all of Europe's wind energy, not all of the properties will be suitable for the installation of a domestic wind turbine. For a wind turbine to be a viable option for use in a domestic installation a minimum annual wind speed of 5m/s will be required, but it will be unlikely that this will be present in an urban area as there will be a lot of buildings around and the wind is likely to be turbulent rather than steady.

Small domestic wind turbines work best in exposed locations. The greater the wind speed the greater the amount of energy which will be produced: an increase in wind speed of just 1m/s or 20% will nearly double the output of most wind turbines. For this reason the correct siting of the turbine is really important. It is also worth considering putting the turbine as high as possible because the wind speed becomes greater with height.

The average mean wind speed for a specific location can be found quite simply on the internet, all that is required is that the site ordinance survey map grid location or the post code is entered. It must be remembered that the values given will only be predicted wind speeds, and these should not be relied on to be completely accurate. A much better method is to monitor the wind speeds over a period of time, as long as three months would be sensible because a wind turbine installation will be expensive. There is little point in installing a system if the amount of energy gained is not going to be enough to pay back the costs of the installation in a reasonable time.

There are many types of wind monitoring systems which will measure and record the average wind speed over a period of time. It is far better to spend a £100 or so on a piece of equipment to satisfy yourself that the thousands of pounds which you are looking to invest is a worthwhile proposition, than to find out after it has been installed

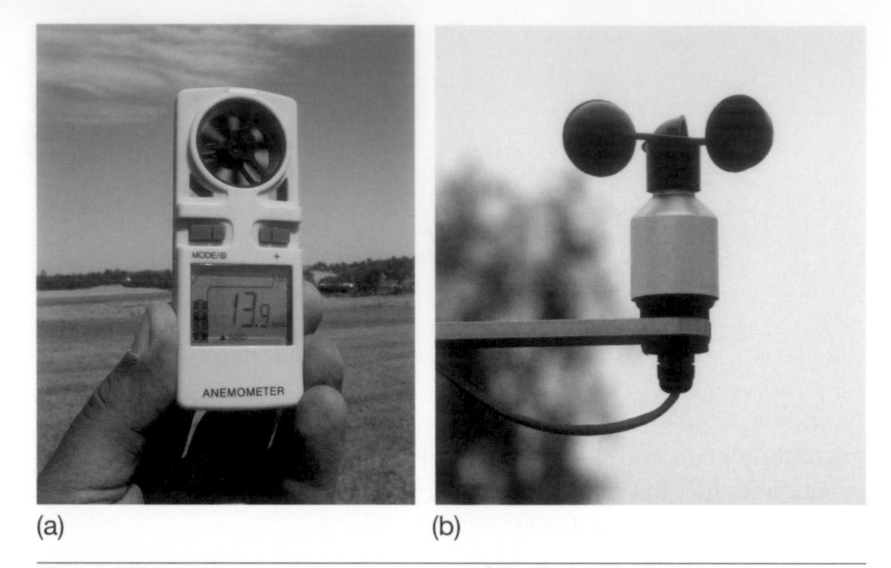

(a) (b)

Figure 3.1 (a) Hand held anemometer for measuring wind speed. (b) Fixed anemometer.

that you have wasted your money. A device which measures wind speed is called an anemometer (Fig. 3.1a/b).

To ensure that a wind turbine is going to be cost effective it is very important that the proposed installation site is suitable and that the installation meets the appropriate standards. The standard for wind turbines is the Microgeneration Installation Standard 3002 and the best practice guide is CE72, both of these documents can be downloaded from the internet.

If a small-scale domestic wind turbine is to be installed to its best advantage it should not be fixed onto a dwelling or any other building as any wind blowing onto a building will cause turbulence which in turn will have a negative effect on the efficiency of the turbine. Although of course once the turbine is installed and paid for any energy produced is free.

A turbine fitted to a tower or mast is by far the best method of installation and the higher the turbine the better it will perform. In general if the required output is to be greater than 1kW the height of the tower would need to be between 15 and 25 metres, this of course depends on site conditions which are very important.

Consideration must be given to whether or not the site is on a hill as wind will speed up as it nears the top of a hill but slows down as it goes down the other side. Clearly wind direction is important, particularly if you are going to install your wind turbine on the side of a hill as it will have a dramatic effect on the amount of energy produced.

Local obstructions such as trees or buildings must also be considered as these again will have an effect on the amount of energy

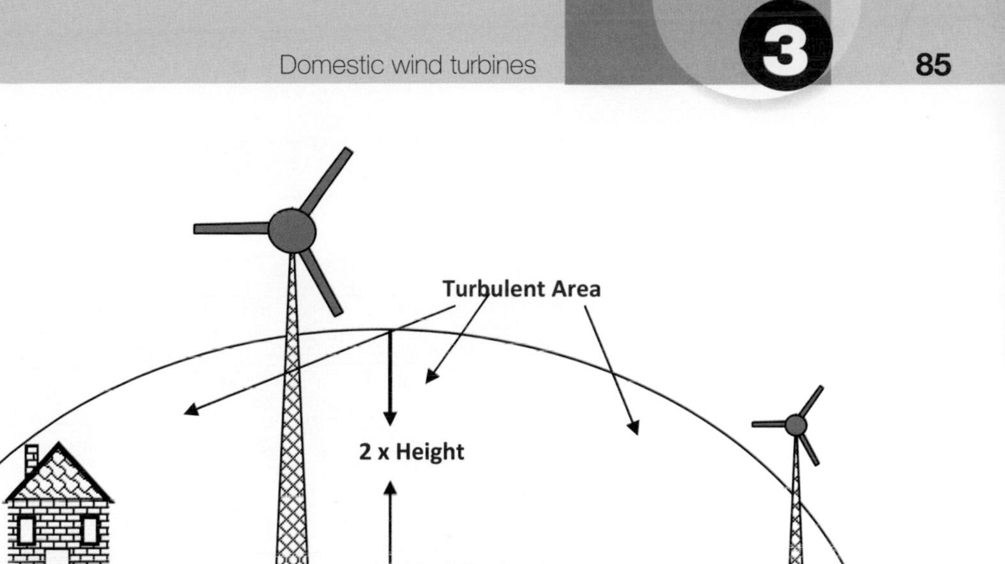

Figure 3.2 Any obstruction will cause turbulence and effect the operation of a wind turbine.

produced. If the wind passes through the turbine before it reaches an obstacle it will not have a detrimental effect on the output of the turbine, unless of course the turbine is very close to the obstacle. However if the turbine is behind the building the wind speed will be greatly reduced which in turn will reduce the output of the turbine to a level which would render it uneconomical. As a rule of thumb a wind turbine sited on a tower which is likely to be downwind of an obstruction should be either twice the height of the obstruction or 20 times the height of the obstruction away from it to ensure that the turbulence caused by it has no effect (Fig. 3.2).

Planning permission

Unlike many other types of renewable energy, micro and small-scale wind turbines will probably require planning permission, and they will certainly require planning permission if they are to be installed in conservation areas. It is important that in all cases the planning authorities are consulted.

In all cases the wind turbine must be sited so that it does not cause disturbances to neighbouring properties. The wind turbine will generate aerodynamic noise due to wind passing over the turbine blades, and also mechanical noise which will be produced by the generator. Larger turbines will also cause a flicker effect which could be very irritating to a neighbour who can see it from a window or from their sitting area in their garden.

Figure 3.3 A fold down mast will provide access for easy maintenance.
Source: Nexgen Wind Monitoring Solutions

Once it has been ascertained that is will be permissible to install a wind turbine, and that a suitable site is available, the type of tower which is going to be used must be selected.

There are various types of mast available which will of course vary in cost and suitability. By far the best type would be a single pole mast which is set directly in the ground, preferably with a mechanism which will allow the mast to fold down for access to the wind turbine (Fig. 3.3).

The added bonus of this type of mast is that it takes up very little space, however the foundation for it would need to be very solid. The mast can be rooted straight into the ground or bolted onto a solid base.

Due to the fact that the mast is going to be positioned to use the wind to its best advantage, it is fairly obvious that it is going to be sited in a windy place. All towers must be installed to be able to withstand very high winds without becoming a hazard, so the tower and the turbine must be installed to be able to withstand winds with an average speed of 78mph (35m/s) for a minimum of ten minutes without any damage. It must also be able to withstand without damage strong gusts of wind of at least 112mph (50m/s).

Every mast should be designed to suit the area in which it is to be situated and be suitable for the type of turbine which it is being used to support.

All fixings must be arranged so that they do not loosen due to vibration and changes in temperature. Special nuts and bolts can be

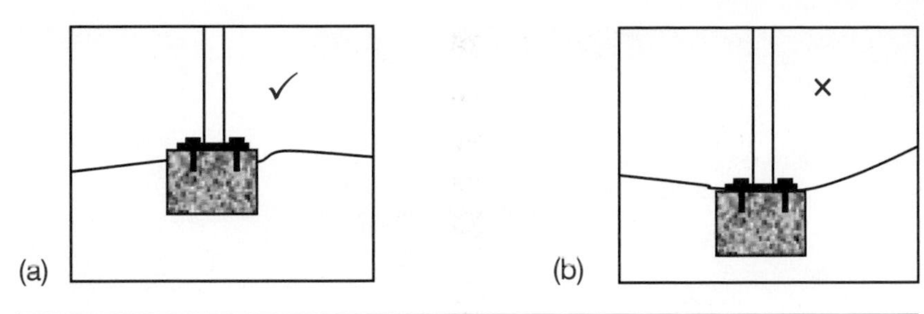

Figure 3.4 (a) The mast fixings should be situated to prevent water collecting around them. (b) This is a poor installation as the bolts and base will corrode quite quickly.

Figure 3.5 Shackles should be provided with a means to prevent them from coming loose.
Source: Nexgen Wind Monitoring Solutions

used or the threads can be treated with a product called 'Loctite', this sets and secures the nuts or bolts in place.

To avoid the need for regular painting the structure should be manufactured from corrosion resistant materials, the structure should also be made to prevent unauthorised climbing.

Any moving parts must be at least three metres from livestock or areas where people may stand. In some instances this will require the installation of a fence.

Depending on site conditions the mast can be either set directly into the ground encased in concrete or it can be bolted to a concrete foundation, but whichever method is used the concrete must be of a suitable strength to suit the foundation requirements. The concrete must also be well compacted around the mast or fixings. Care must be taken to ensure that water or other liquids cannot pool around the base of the mast or the anchor points (Fig. 3.4a/b).

Where a single pole mast is used without a good foundation it would be necessary to support the mast using guy lines. This of course will result in the mast having a larger footprint as the guy lines will need to extend over quite a large area. Where guy lines are used it is a requirement that any shackles or other cable fixings must be provided with a means to prevent them from loosening (Fig. 3.5).

If the wind turbine is to be installed in a remote area where the visual impact of it is not important then a lattice type tower (Fig. 3.6) can be used. These look very similar to electricity pylons and must also be constructed to withstand the same wind speeds as a mast, they must also be designed to prevent persons from climbing them.

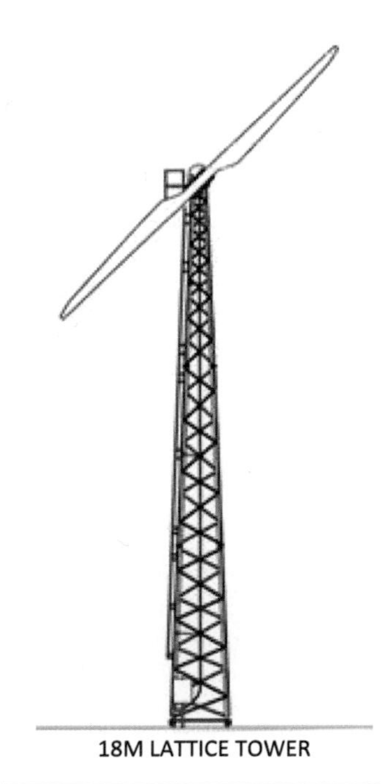

Figure 3.6 Lattice towers can be unsightly and it is sometimes difficult to prevent them being climbed.

Where a mast or tower is installed near to a building which has its own lightning protection system, consideration must be given as to whether the mast or tower needs to be electrically connected to the lightning system or not. The best way to deal with this is to consult with a lightning protection specialist as it is not really a decision

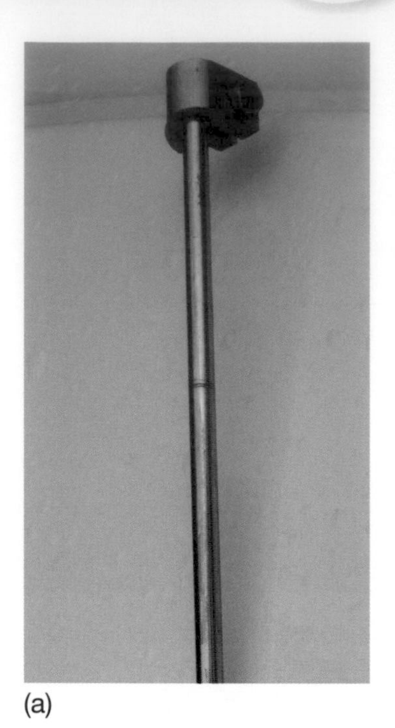

(a)

(b)

Figure 3.7 (a) A typical earth electrode and termination. (b) The electrode should be sited as close to the structure as possible.
Source: Wikipedia

which can be made without good knowledge of the requirements of BS EN 62305 for lightning protection.

Consideration must also be given as to whether the structure needs to be bonded to the main earthing terminal of the electrical installation to which it is being connected. Bonding would not be required unless the structure is within the equipotential zone of the building.

Where the structure is not connected to a lightning protection system or equipotential bonding, an earth electrode must be provided as close as possible to the structure (Fig. 3.7a/b).

The earth electrode must not have a resistance of more than 10Ω and must be connected to the structure by a copper conductor with a minimum CSA of $16mm^2$. The conductor must not have any sharp bends and must be clearly identified. Protection must also be provided for the earth electrode and enclosures are obtainable specifically for this purpose (Fig. 3.8).

Figure 3.8 An electrode must be provided with protection and enclosures are made for this purpose.
Source: MK Electric

Measurement of the earth electrode resistance is carried out using an earth electrode test instrument (Fig. 3.9).

There are various types earth electrode test equipment, most of which require a resistance value to be measured at three different points and an average of the three values to be taken as the electrode resistance (Fig. 3.10).

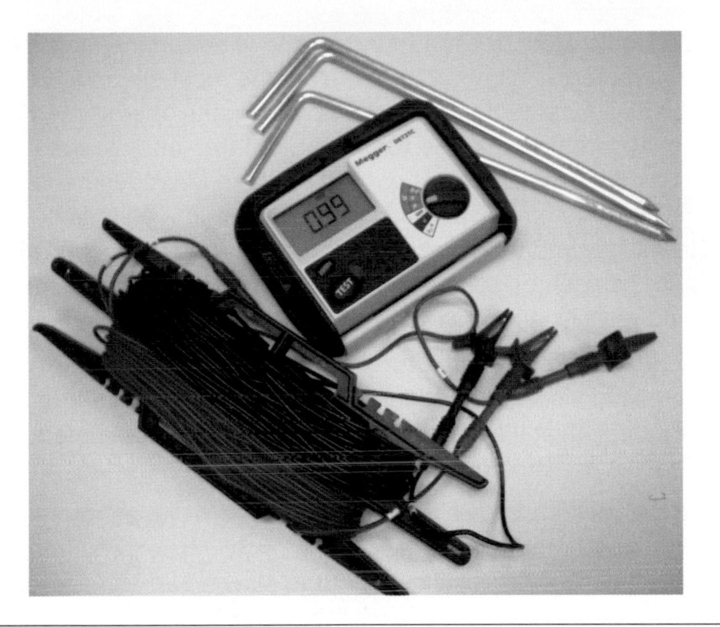

Figure 3.9 Earth electrode test instrument.

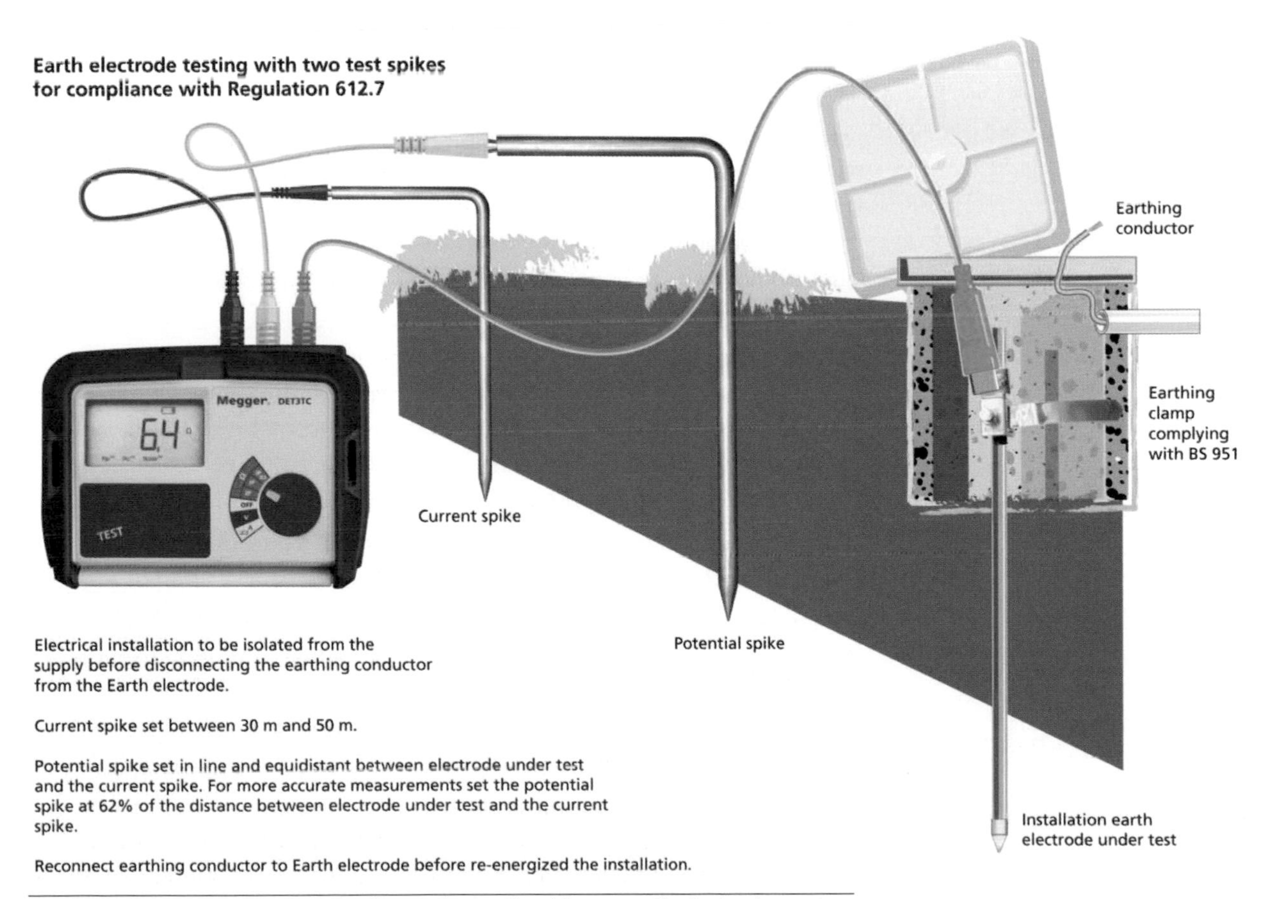

Figure 3.10 This drawing shows the testing method for an earth electrode.

Once a suitable site has been found for the wind turbine, and a mast has been erected, it is important to select the most suitable type of turbine. If it is going to be used to supplement the energy used from the national grid the larger the turbine output the better the results. Remember, if the installation is carried out by a Microgeneration Certification Scheme (MCS) registered installer your electricity supplier will pay you for every kW generated, and any energy which you do not use can be sold to your energy supplier at a higher rate than if you were using it yourself.

Where it is at all likely that the turbine will have to contend with high winds, precautions have to be taken to ensure that the turbine blades do not rotate so fast that the generator burns out and the blades start to fall apart. A common method is to install a furling device.

A furling device (Fig. 3.11) ensures that when the blades rotate at a predetermined wind speed, they automatically turn away from the wind, this of course will allow them to slow down and prevent any damage. Another method is for the owner to simply turn the installation off for a while and manually turn the blades away from the wind, but a furling device is a far better option as it will allow the turbine to keep generating even in high winds.

Wind turbines come in various sizes and ratings, so it is important to select one with an output which will be suitable for the installation requirements. To be able to design the installation correctly we must know the maximum voltage $V_{(max)}$ and current output $I_{(max)}$ which the turbine will produce at various wind speeds up to a maximum of 50m/s. We should also know the cut-in wind speed of the turbine, just above 3m/s is usual, at wind speeds less than the cut-in speed the turbine will not work at all.

Small-scale wind turbines generally produce alternating current (a.c.) at between 12 volts and 48 volts, and can be either single phase or three phase. D.c. wind turbines are sometimes used, but unfortunately because of the way they work they have a commutator and brushes which require maintenance (Fig. 3.12).

D.c. turbines are most common on very small systems which are used for trickle charging batteries.

Where a turbine is used which produces a.c. it is usual for the current to be converted to d.c. by using a rectifier either within the turbine or positioned between the turbine and the inverter.

The current which we use in our houses is 230 volts a.c. at a frequency of 50Hz. The wind turbine will produce a voltage and frequency which will be dependent on the wind speed, but this of course will change constantly on most days. Because we want to use our free electricity in our homes it will have to be connected to our supply source, but to enable this to happen the energy produced

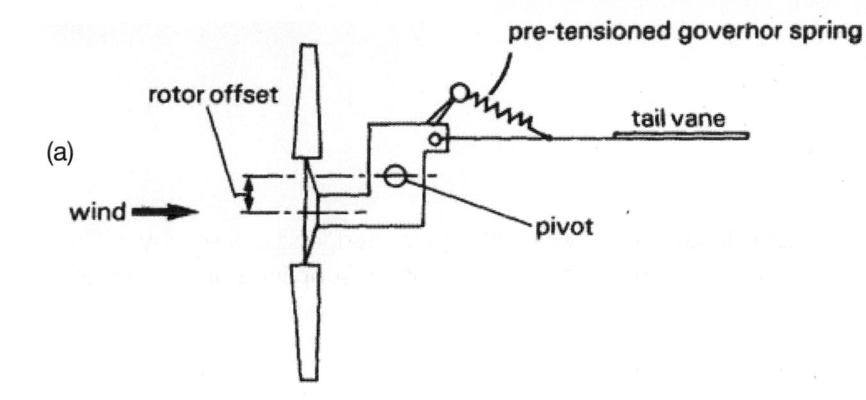

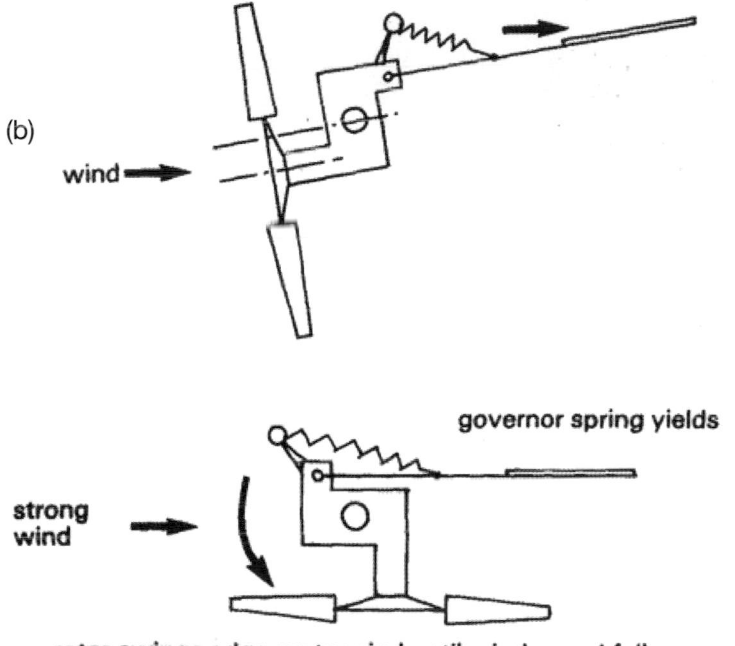

Figure 3.11 Too much wind could result in damage to the wind turbine, so a furling device is used to turn the blades away from the wind when it reaches very high speeds.

must be at the same voltage and frequency as our supply. This requires the energy from the turbine to be fed into a device called an inverter. It is important that the inverter is matched to the output of the wind turbine being used as if it is not the efficiency of the inverter will be affected and in some cases it may not even work at all. An inverter (Fig. 3.13) is a device which takes direct current at a voltage within its operating range and synchronises it with the supply voltage and frequency, allowing it to be directly connected to the main building supply.

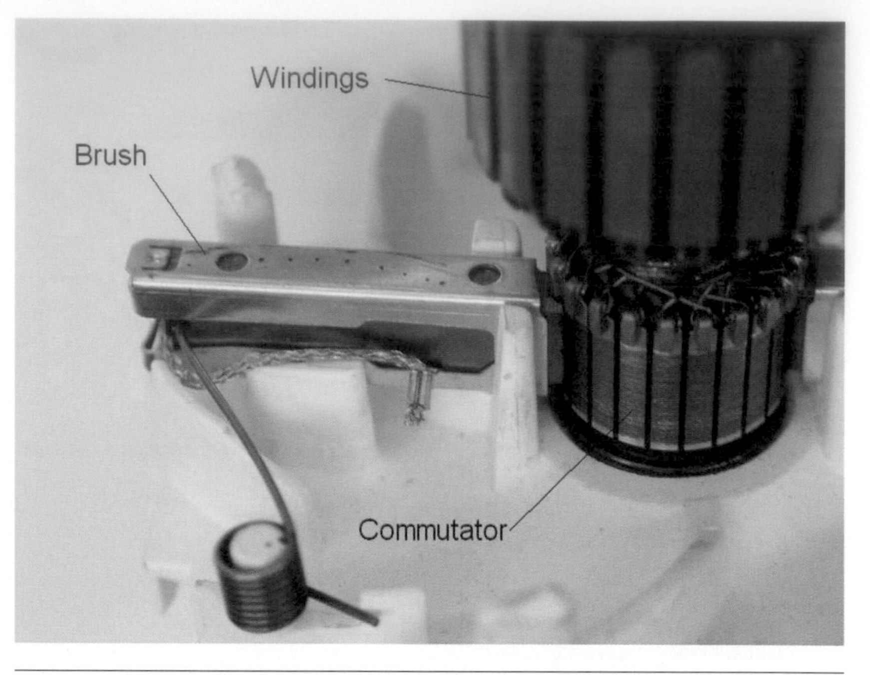

Figure 3.12 A commutator and brushes will require maintenance.

Figure 3.13 An inverter will convert direct current into alternating current and synchronise it with the supply voltage and frequency.

The positioning of the inverter will be a matter of choice, although site conditions may determine where it is to be installed. The inverter can be positioned near to the turbine if required although it is more usual for it to be near the consumer's unit at the origin of the main supply.

A cable is used to transport the output from the turbine to the building in which the energy is to be used and the most basic requirement of course is that the cable is suitably rated to carry the maximum voltage $V_{(max)}$ and current $I_{(max)}$ which the turbine could produce at between 0-50m/s. There is no requirement for the output cable to be protected against overload or fault current, providing the cable is of a suitable size, this is because the turbine will not be able to produce a voltage or a current at greater than its maximum. The cable must also be protected to deal with any environmental conditions which may affect it along its entire route.

A suitable cable for use with wind turbine systems would be steel wired armoured cable, which apart from being suitable for the maximum voltage and current must also comply with all of the other requirements of BS 7671 wiring regulations 2008. Information on cable calculation can be found in appendix 4 of BS 7671.

Consideration must be given to whether the cable is grouped with cables of other circuits, and allowances must be made for temperature and any insulation which the cable may pass through. If the cable is to be buried it will result in the cable current carrying capacity being reduced by at least 10%.

Volt drop is also something which needs to be considered as any loss of voltage will result in a loss of energy. It is far more cost effective to install a larger cable to keep volt drop to a minimum than to install a smaller (cheaper) cable and lose energy over a long period of time. The recommended maximum volt drop for a cable from a wind turbine is 4%, but this is only a recommendation and not a requirement. However, as previously mentioned, it is far better to keep the voltage drop as low as possible as this will increase the installation's efficiency. The calculation for voltage drop is quite a simple process providing the required information is available.

As an example let us assume that the inverter is to be positioned in the house near to the metre position which is 43m away from the wind turbine. The turbine is producing 24 volts and a current of 3.2 amps d.c. Voltage drop must be no more than 4% and the cable is steel wire armour.

The maximum permissible volt drop can be found:

$$\frac{24 \times 4}{100} = 0.96\text{v.}$$

Now we can work out the maximum volt drop per metre which would be permitted for our cable.

$$\frac{\text{permitted volt drop} \times 1000}{\text{amps} \times \text{length}} = \text{mv/A/m}$$

$$\frac{0.96 \times 1000}{3.2 \times 43} = 6.97\text{mv/A/m.}$$

If we assume that the cable being used is steel wire armoured then we need to look in table 4D4B of BS 7671. We can see that a 10mm^2 two core d.c. cable has a volt drop of 4.4mv/A/m which is far less than our maximum permitted.

The actual volt drop would be:

$$\frac{\text{millivolts} \times \text{amps} \times \text{length}}{1000} \text{ actual volt drop}$$

$$\frac{4.4 \times 3.2 \times 43}{1000} = 0.6 \text{ volts which is only } 2.5\%.$$

and as we can see from column 6 of table 4D4A the current rating for a 10mm^2 cable buried in the ground is 60A.

Of course if the inverter is installed at the turbine end then a similar calculation must be carried out for volt drop, the difference being that the permissible volt drop is now limited to 1% of the output voltage of the inverter. A volt drop of greater than this will result in the protection system for the inverter being compromised.

Once the cable has been installed it must be connected to an isolator at the turbine end which must be capable of disconnecting all of the conductors and if the current has been rectified to d.c. then the isolator must be d.c. rated (Fig. 3.14). This is because d.c. current creates a much larger arc when the switch contacts part and this would have a damaging effect on an a.c. isolator.

Figure 3.14 A d.c. rated isolator must be used on the d.c. side of the system.

Any isolator used must be matched to the turbine and be suitable for the voltage and current which will be produced. The conductors must be correctly identified and if they are d.c. they must be colour coded brown for positive and grey for negative. The information for colour identification can be found in appendix 7 of BS 7671.

The isolator must be clearly labelled (Fig. 3.15) so as to ensure that anybody who is likely to have to work on the installation is well aware that the isolator is for a wind turbine.

On and off position must be clearly indicated and the isolator must be able to be secured in the off position.

A device for locking the turbine itself and preventing it from rotating would not be accepted as a suitable means of isolation.

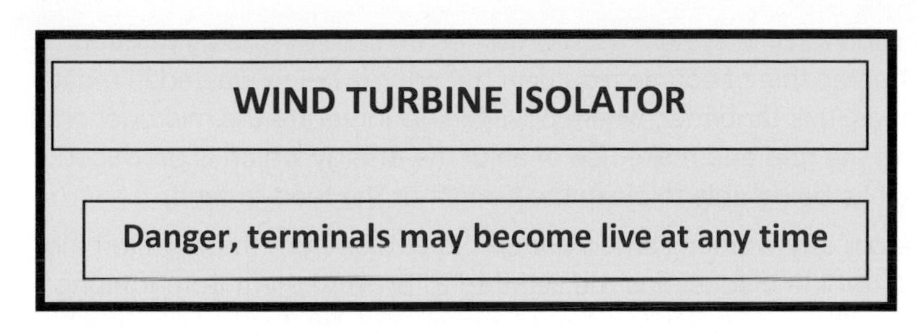

Figure 3.15 Clear identification of equipment is very important.

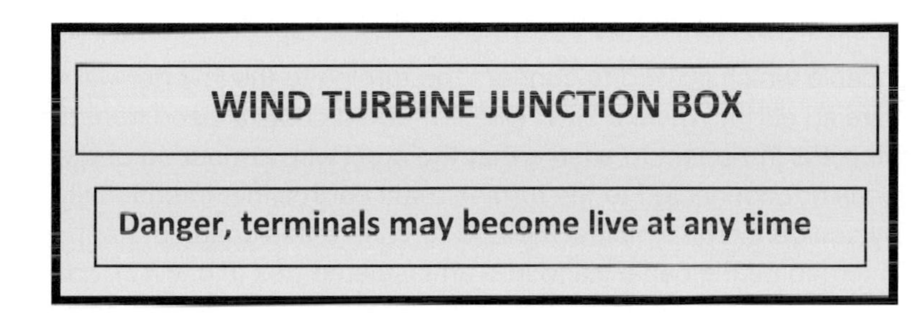

Figure 3.16 Junction boxes must also be labelled.

If a junction box is going to be used to join the output cable it must also be labelled clearly, see (Fig. 3.16).

A junction box situated in the right place can be very useful as a test point for the turbine.

Connection of the wind turbine to its point of use

Wind turbines can be used as a directly connected system and in this type of system the energy produced by the turbine is used by the load which it is connected to. For practical reasons this type of load would perhaps be a water heating system. Obviously the system would only work on windy days and for this reason would only be used to supplement the main supply. The load would normally be a part of the installation which is dedicated to the supply and would be fed through a controller. To save losses the load would be d.c.

In many ways this type of system operates in the same way as a solar thermal system: when renewable energy is available it is used to save the energy which would have been used from fossil fuels.

The problem with this type of system is that when the load is satisfied, and energy is not required, anything which is generated has to be dumped.

A wind turbine system usually dumps its surplus energy through a heater, this of course results in the energy being wasted in most cases. It is far better, where possible, to integrate the microgeneration with the grid and make use of all of the energy which is produced, as well as being able to claim the benefit of the feed-in tariff.

Before any work in carried out on the cables from the wind turbine the turbine blades must be secured to prevent them from rotating. Once the cables are connected into an isolator, the isolator must be secured in the open position. Failure to do this could result in electric shock. Relying on the turbine lock is not accepted as a safe means of isolation when working on the system.

Any cable which is used to connect the turbine to the inverter will not require an earth. Where a steel wire armoured cable is used from the turbine it is important to ensure that the steel wire armouring of the cable is not connected to the turbine mast earth either intentionally or by accident. The simplest method to ensure this does not happen is to terminate the cable gland into an insulated box (Fig. 3.17), and it is a good idea to do this at both ends of the cable. The wire armour is only used as mechanical protection and there is no requirement

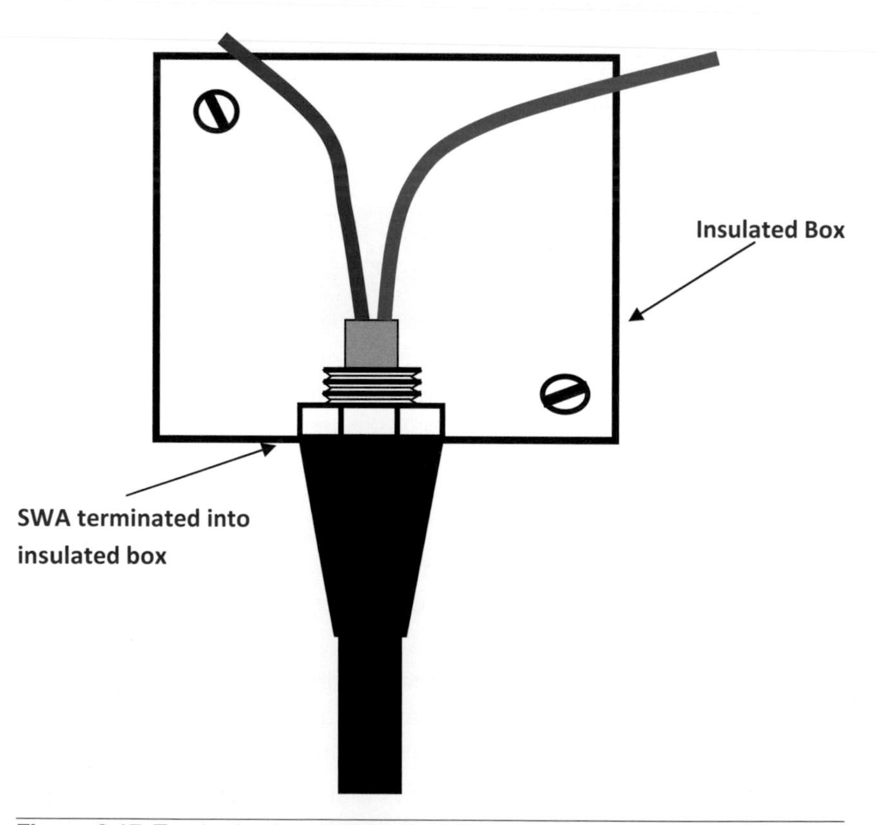

Figure 3.17 Termination into an insulated box will prevent the transportation of an earth into an area where it could cause a danger.

for it to be connected to earth. If it was connected to the mast and the mast was stuck by lightning the damage to the fixed installation would be massive.

At the load end of the cable, before the cable is connected to the inverter, it must have a d.c. switch disconnector fitted to enable the inverter to be isolated from the d.c. supply from the turbine. On the output side of the inverter an a.c. isolator must be installed to enable the inverter to be isolated from the low voltage part of the installation. Both isolators must be correctly labelled indicating exactly what their purpose is.

Where the inverter is not adjacent to the point of connection to the low voltage installation it must have an isolator next to the inverter and another at the point of connection.

The selection of the inverter is very important. As previously mentioned, if it is not matched to the efficiency of the system it will be dramatically reduced and even worse, it may not operate at all.

All inverters must have a test certificate as recommended in the engineering recommendation G83/1 which is a document which sets out the requirements for microgeneration systems up to 16 amps, above this engineering recommendation G59/1 must be used.

The certificate provided with the inverter will provide information on the following:

- power quality
- over and under frequency switch off
- over and under voltage switch off
- loss of mains test
- reconnection time measurement.

This information is required to prove that the inverter meets the standards required.

Where inverters are used to the standard required by G83/1 they must have an automatic protection system to include:

- When there is a loss of the main supply the inverter must have a drop out system which automatically shuts the inverter down, this is required to prevent the installation being fed by the turbine when it may not be expected. Where the system will be expected to sustain some circuits in the event of a power failure a battery backup system must be used.

Imagine the consequences if the microgeneration system, in this case the wind turbine, fed energy back into the supply system when the supply provider had shut down the supply to carry out repair or maintenance work, or even during a power cut.

When a microgeneration system back feeds into a supply system under these circumstances the condition is known as islanding, and this would be a real danger to any persons working on the supply system. An inverter which has been constructed to the requirements of G83/1 will have all of the required safety features.

The inverter must also stop operating when:

- The voltage rises above 264v or falls below 207v.
- The supply frequency rises above 50.5 Hz or falls below 47 Hz.

On the load side of the inverter an a.c. isolator must be installed (in some documents an isolator is referred to as a switch disconnector). This isolator, along with the d.c. isolator provided on the turbine side of the inverter, will provide safe isolation for the inverter and allow any required work to be carried out on it safely. If the inverter is a distance from the point of connection to the incoming supply then two a.c. isolators will be required, one next to the inverter and another as close as possible to the point of connection to the supply.

Where the inverter is in the same room as the point of connection it is only a requirement for one isolator to be installed, but in all cases the isolator must be compliant with BS EN 60947-3.

Isolators must also:

- be clearly marked to indicate on and off positions
- be lockable in the off position only
- switch all live conductors, this would of course include the neutral conductor
- be clearly labelled as shown in Figure 3.18.

WIND SYSTEM Point of emergency isolation

Figure 3.18 Labels and notices must be very clear.

Interconnection to the low voltage supply

When connecting the microgeneration system to the fixed wiring of the installation the first consideration is to ensure that the supply from the wind turbine has been correctly isolated, it is also a very good idea to isolate the a.c. side of the inverter as well.

Connection to the supply system can be by various methods and can be made through a fuse combination unit or directly into the consumer's unit through a dedicated circuit using a spare protective device (Fig. 3.19a/b).

Where the interconnection is made through a fuse combination unit the connection from the unit to the installation can be made by using what some call a Henley Block or a mains connector block (Fig. 3.20).

The short circuit current on the load side of the inverter will be equal to the load current, this is because it is the maximum current which the wind turbine can produce. For this reason short circuit protection

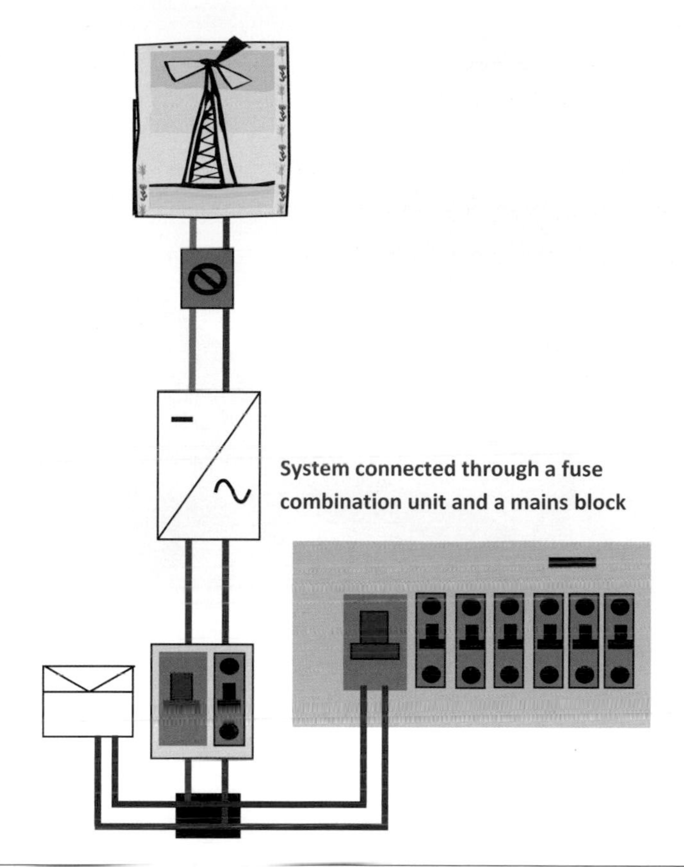

Figure 3.19 (a) Connection through a fuse combination unit.

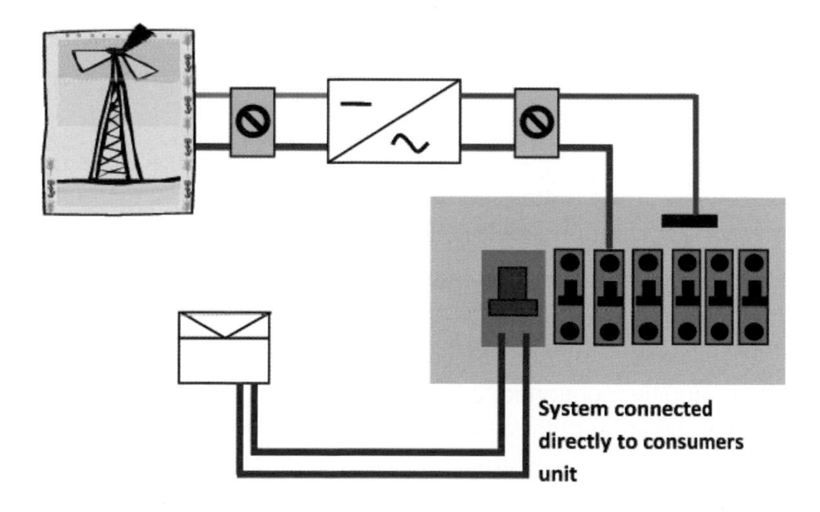

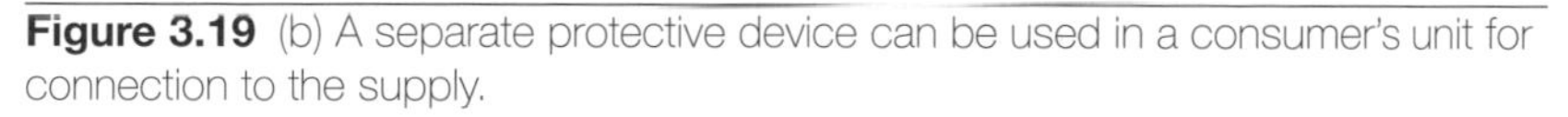

Figure 3.19 (b) A separate protective device can be used in a consumer's unit for connection to the supply.

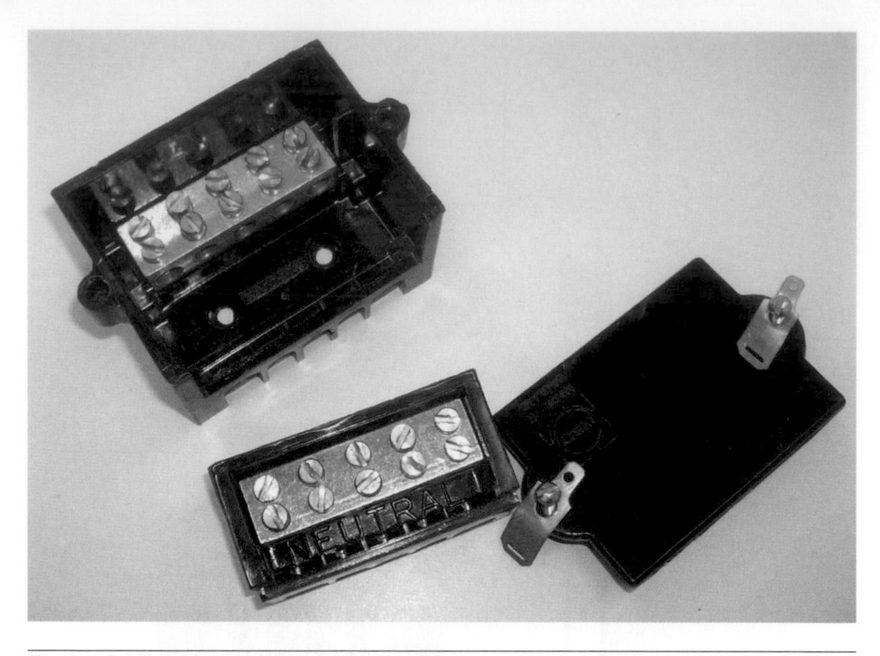

Figure 3.20 A Henley Block which could be used to connect to existing metre tails.

on the inverter load side is unnecessary and would not provide any additional safety. However, the cables from the installation supply to the inverter must have protection to comply with BS 7671. The protective device fitted in the consumer's unit or fused combination unit would provide the necessary protection, but of course care must be taken to ensure that the PFC rating of the protective device is suitable for the installation. In other words it must be able to withstand the highest fault current which could flow in the installation.

Providing the inverter is compliant with G83/1h, rcd protection at the point of interconnection is not required.

Where the wind turbine is being used as a stand alone system or with battery backup a battery controller should be installed using the same method as described in the chapter on photovoltaics.

Due to the fact that an installation with a wind turbine system connected has two systems of supply, notices and labels are a very important part of the installation.

As previously mentioned, all conductors must be clearly identified, as must all isolation points. In addition to this a notice clearly showing that the installation includes on-site generation must be provided to indicate where the isolation points for the main supply and the on-site generation are situated (Fig. 3.21).

Also at the point of interconnection it is a requirement that the following information is provided:

- a circuit diagram

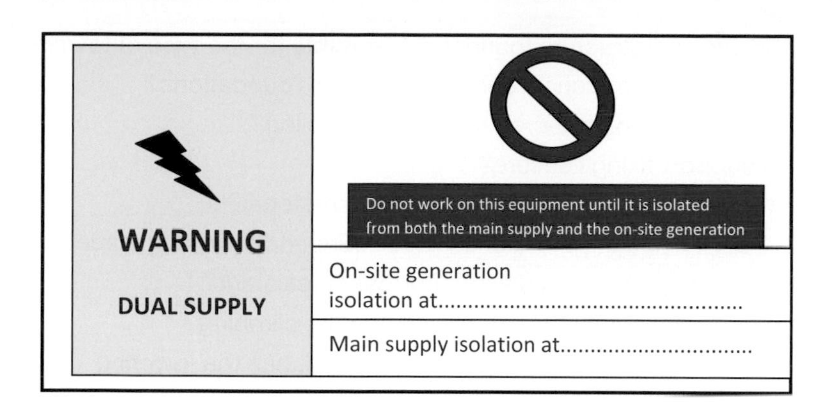

Figure 3.21 A sign indicating that the system has two points of isolation must be provided at each isolation point.

- a contact number for the installer/ supplier/any person who maintains the system
- a summary of the protection settings which are part of the system.

Commissioning and testing

Small wind turbines are no different than any other type of electrical system when it comes to the commissioning of the installation. An initial verification must be carried out which includes a visual inspection followed by very thorough testing, and documentation must be completed to verify that the installation is safe and fit for use.

Without the required documentation access to a microgeneration installers scheme will not be permitted, this in turn will result in the person who owns the installation being unable to sign up with the distribution network operator and benefit from the feed-in tariff.

The visual inspection must include the following.

Description of the system

For example:

The type of turbine and its rating, is it grid connected or is it a stand alone system?

If it is a stand alone system, what is it being used for?

Turbine siting

Is the turbine installed in a safe location?

Are exposed moving parts a minimum of 3m away from persons or livestock?

Has the support structure been chosen to suit the maximum wind load?

Are the support structure foundations suitable for the type of tower?

Have the correct materials been used for the foundations?

Are the foundation protected from water pooling?

Are all support fixings secure?

If guy lines are used are they of a suitable material?

Are all guy turnbuckles and shackles secure and locked in place?

Are the guy anchors suitable and corrosion resistant?

Has the turbine mast been protected against climbing?

If the mast is of a type which can be lowered, has the lowering equipment been safely stored?

Turbine installation

Are the manufacturers installation instructions available and has the turbine been installed in accordance with them?

Is the turbine braking system working correctly?

Are vibration levels acceptable at all turbine speeds?

Output cables

Are cable sizing calculations available?

Is the volt drop suitable for the installation, if greater than 4% will this be acceptable?

Are the cables of a suitable current rating and suitable for the method which has been used for the installation?

Has care been taken with the installation of the cables, are they correctly secured and at a suitable distance from any external influences which may cause them damage?

Are the electrical connections to the turbine secure and protected from the environment?

Turbine isolator

Is the turbine isolator correctly rated (d.c./a.c.)?

Has it been correctly installed in the correct place with secure terminations?

Is it labelled correctly?

If a turbine junction box has been used, are the electrical connections secure, is it in a suitable location, has the IP rating of the junction box been documented?

Earthing

Is the turbine mast earthed correctly?

Is surge protection required? If so is it correctly installed?

Metering

Is the metre to the required BS?

Is it sealed?

Are the connections tight?

Has the system been explained to the customer?

Turbine controller

Is the voltage and current rating and of the unit correct?

Is the controller installed correctly with all electrical connections secure?

Is the IP rating for the controller suitable for the environment and is the IP rating documented?

Is the controller suitably ventilated and mounted to prevent damage due to heat transfer?

Is it labelled correctly?

For compliance with the microgeneration certification scheme all systems have to have an operation and maintenance manual produced which is specific to each individual installation. This manual must be given to the customer on completion of the job after all inspections and testing has been completed and the installation is considered to be safe and suitable for use.

Each manual must contain as much helpful information as possible, in all cases it must contain the following:

- Contact details of the installer.
- Documented procedures for starting and stopping the turbine.
- If it is possible to lower the mast, information must be provided on how to lower and raise it safely.
- All manufacturers' information/manuals for the turbine, inverter, battery controller, etc.
- All warranty details for equipment.
- Completed inspection and test schedules.
- Maintenance details, including information on lubrication if required.
- Recommended time between inspections.
- A maintenance schedule/record sheet.
- A flow chart to help with any fault diagnostics.
- Information on system design, this is to include schematic diagrams, $V_{(max)}$, $I_{(max)}$, electrical drawings, cable routes and foundation details.

If the wind turbine system is connected to the grid then the following documentation will also be required:

- An electrical installation certificate to BS 7671.
- Signed approval from the network operator to show that they will accept the installation.
- A print out of the protection settings (this will include the certificate provided with the inverter).

Direct connected systems

The following items have to be inspected:

Is the load suitable?

Is the turbine suited to the load, with regards to load current and voltage?

Has the correct controller been installed?

Where dump heaters are used are they positioned to prevent fire, is the cable used heat resistant?

Is the labelling complete?

Is all documentation in place?

Battery connected systems

Is it possible to manually isolate the battery?

Is battery over-current protection provided?

Are the correct batteries fitted for use with the type of system (load and charge rate)?

When the battery is isolated it must not be possible for the turbine to supply the load directly.

Are the batteries in a ventilated location or of the glass matt or gel type which do not require ventilation?

Are the cables correctly routed, rated and protected by d.c. fuses?

Where an inverter is used is it suitably mounted (heat/ventilation)? Can it be safely isolated?

Are all metres correctly installed and rated?

Where the system supports a battery system and is integrated with the grid, or if the battery system supplies an a.c. load through an inverter, all of the a.c. side of the installation must be installed to the requirements of BS 7671.

Grid connected systems

Does the a.c. side of the installation comply with BS 7671?

Is it possible for the inverter to be isolated for maintenance?

Are all isolators lockable in the off position?

Is the inverter correctly rated and installed for heat dissipation?

Are all notices in place?

Has the inverter been checked for complete disconnection should the supply fail?

Is all documentation in place?

As well as a checklist showing all of the items which have been inspected, the a.c. side of the installation must have an electrical installation certificate, along with a schedule of inspections and a schedule of test results (Figs. 3.22–3.24). This documentation must be specific, and only include information for the part of the installation which relates to the renewable energy and the origin of the main supply. This information must not be added to existing certification.

Once it has been verified that the installation is complete, and safe to test, it is very important that for safety reasons testing is carried out in the correct sequence.

For the wind turbine (microgeneration) part of the installation the testing is minimal, this is of course providing the inspection has been satisfactory. Most of the information required to prove correct operation will be gained by reading the digital display metre when the turbine is operating.

One test that would be required is that the earth electrode at the turbine mast has a resistance of no more than 10Ω, this of course is really just a check as the electrode resistance should have been verified when it was first installed. This would require the use of an earth electrode test instrument.

On the grid side of the installation testing must be carried out from the inverter to the point of integration, this testing would be just as it would be for any other low voltage installation. The sequence is:

- continuity of protective conductors
- insulation resistance
- polarity
- earth fault loop impedance
- prospective short circuit current
- functional testing
- recd if installed.

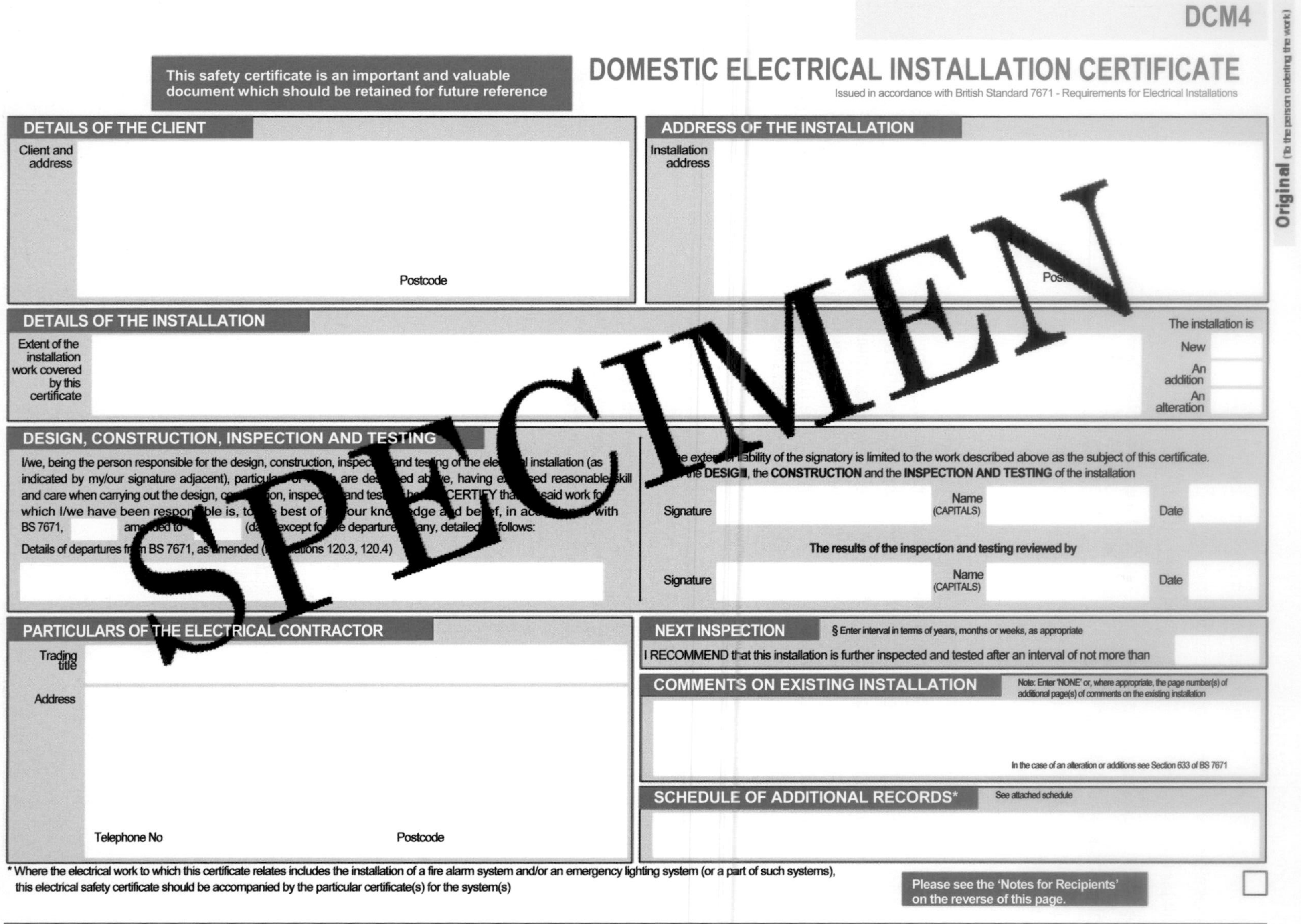

Original (To the person ordering the work)

DCM4

This safety certificate is an important and valuable document which should be retained for future reference

DOMESTIC ELECTRICAL INSTALLATION CERTIFICATE

Issued in accordance with British Standard 7671 - Requirements for Electrical Installations

DETAILS OF THE CLIENT

Client and address

Postcode

ADDRESS OF THE INSTALLATION

Installation address

Pos[illegible]

DETAILS OF THE INSTALLATION

Extent of the installation work covered by this certificate

The installation is: New / An addition / An alteration

DESIGN, CONSTRUCTION, INSPECTION AND TESTING

I/we, being the person responsible for the design, construction, inspec[illegible] and testing of the ele[illegible] installation (as indicated by my/our signature adjacent), particula[illegible] are de[illegible]ed ab[illegible]e, having e[illegible]sed reasonable skill and care when carrying out the design, c[illegible]on, inspec[illegible] and tes[illegible] CERTIFY tha[illegible] said work fo[illegible] which I/we have been respon[illegible]ble is, to [illegible] best of [illegible] our kno[illegible]dge a[illegible]d be[illegible]ef, in ac[illegible] with BS 7671, ______ am[illegible]ed to ______ (da[illegible] except fo[illegible] departure[illegible] any, detailed [illegible] follows:

Details of departures fr[illegible]m BS 7671, as amended ([illegible]ations 120.3, 120.4)

[illegible]e exte[illegible] liability of the signatory is limited to the work described above as the subject of this certificate. [illegible] the **DESIGN**, the **CONSTRUCTION** and the **INSPECTION AND TESTING** of the installation

Signature | Name (CAPITALS) | Date

The results of the inspection and testing reviewed by

Signature | Name (CAPITALS) | Date

PARTICULARS OF THE ELECTRICAL CONTRACTOR

Trading title

Address

Telephone No | Postcode

NEXT INSPECTION § Enter interval in terms of years, months or weeks, as appropriate

I RECOMMEND that this installation is further inspected and tested after an interval of not more than

COMMENTS ON EXISTING INSTALLATION Note: Enter 'NONE' or, where appropriate, the page number(s) of additional page(s) of comments on the existing installation

In the case of an alteration or additions see Section 633 of BS 7671

SCHEDULE OF ADDITIONAL RECORDS* See attached schedule

* Where the electrical work to which this certificate relates includes the installation of a fire alarm system and/or an emergency lighting system (or a part of such systems), this electrical safety certificate should be accompanied by the particular certificate(s) for the system(s)

Please see the 'Notes for Recipients' on the reverse of this page.

SPECIMEN

Figure 3.22 Specimen Electrical Installation Certificate.

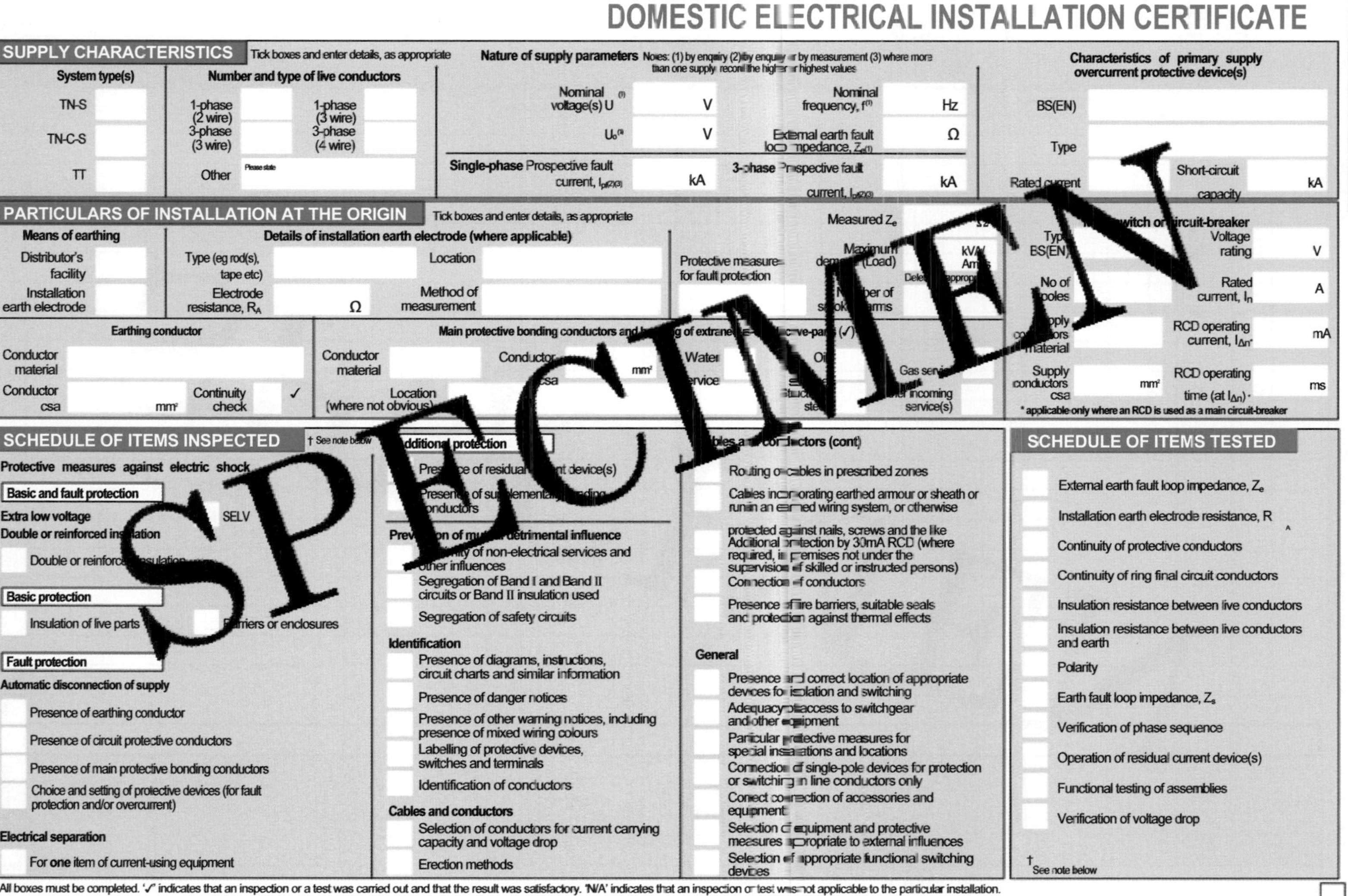

DCM4

Original (to the person ordering the work)

DOMESTIC ELECTRICAL INSTALLATION CERTIFICATE

SPECIMEN

SUPPLY CHARACTERISTICS

Tick boxes and enter details, as appropriate

System type(s)
- TN-S
- TN-C-S
- TT

Number and type of live conductors
- 1-phase (2 wire)
- 3-phase (3 wire)
- 1-phase (3 wire)
- 3-phase (4 wire)
- Other (Please state)

Nature of supply parameters Notes: (1) by enquiry (2) by enquiry or by measurement (3) where more than one supply record the higher or highest values

- Nominal voltage(s) U (1) ... V
- U_0 (1) ... V
- Nominal frequency, f (1) ... Hz
- External earth fault loop impedance, Z_e (2) ... Ω
- Single-phase Prospective fault current, I_{pf} (2)(3) ... kA
- 3-phase Prospective fault current, I_{pf} (2)(3) ... kA

Characteristics of primary supply overcurrent protective device(s)
- BS(EN)
- Type
- Rated current
- Short-circuit capacity ... kA

PARTICULARS OF INSTALLATION AT THE ORIGIN

Tick boxes and enter details, as appropriate

Means of earthing
- Distributor's facility
- Installation earth electrode

Details of installation earth electrode (where applicable)
- Type (eg rod(s), tape etc)
- Location
- Electrode resistance, R_A ... Ω
- Method of measurement

- Measured Z_e ... Ω
- Protective measure for fault protection
- Maximum demand (Load) ... kVA / Amps (Delete as appropriate)
- Number of smoke alarms

Main switch or circuit-breaker
- Type BS(EN)
- No of poles
- Supply conductors material
- Supply conductors csa ... mm²
- Voltage rating ... V
- Rated current, I_n ... A
- RCD operating current, $I_{\Delta n}$* ... mA
- RCD operating time (at $I_{\Delta n}$)* ... ms

* applicable only where an RCD is used as a main circuit-breaker

Earthing conductor
- Conductor material
- Conductor csa ... mm²
- Continuity check ✓

Main protective bonding conductors and bonding of extraneous-conductive-parts (✓)
- Conductor material
- Conductor csa ... mm²
- Location (where not obvious)
- Water service
- Oil service
- Gas service
- Structural steel
- Other incoming service(s)

SCHEDULE OF ITEMS INSPECTED

† See note below

Protective measures against electric shock

Basic and fault protection

Extra low voltage
- SELV

Double or reinforced insulation
- Double or reinforced insulation

Basic protection
- Insulation of live parts
- Barriers or enclosures

Fault protection

Automatic disconnection of supply
- Presence of earthing conductor
- Presence of circuit protective conductors
- Presence of main protective bonding conductors
- Choice and setting of protective devices (for fault protection and/or overcurrent)

Electrical separation
- For **one** item of current-using equipment

Additional protection
- Presence of residual current device(s)
- Presence of supplementary bonding conductors

Prevention of mutual detrimental influence
- Proximity of non-electrical services and other influences
- Segregation of Band I and Band II circuits or Band II insulation used
- Segregation of safety circuits

Identification
- Presence of diagrams, instructions, circuit charts and similar information
- Presence of danger notices
- Presence of other warning notices, including presence of mixed wiring colours
- Labelling of protective devices, switches and terminals
- Identification of conductors

Cables and conductors
- Selection of conductors for current carrying capacity and voltage drop
- Erection methods

Cables and conductors (cont)
- Routing of cables in prescribed zones
- Cables incorporating earthed armour or sheath or run in an earthed wiring system, or otherwise protected against nails, screws and the like
- Additional protection by 30mA RCD (where required, in premises not under the supervision of skilled or instructed persons)
- Connection of conductors
- Presence of fire barriers, suitable seals and protection against thermal effects

General
- Presence and correct location of appropriate devices for isolation and switching
- Adequacy of access to switchgear and other equipment
- Particular protective measures for special installations and locations
- Connection of single-pole devices for protection or switching in line conductors only
- Correct connection of accessories and equipment
- Selection of equipment and protective measures appropriate to external influences
- Selection of appropriate functional switching devices

SCHEDULE OF ITEMS TESTED

- External earth fault loop impedance, Z_e
- Installation earth electrode resistance, R
- Continuity of protective conductors
- Continuity of ring final circuit conductors
- Insulation resistance between live conductors
- Insulation resistance between live conductors and earth
- Polarity
- Earth fault loop impedance, Z_s
- Verification of phase sequence
- Operation of residual current device(s)
- Functional testing of assemblies
- Verification of voltage drop

† See note below

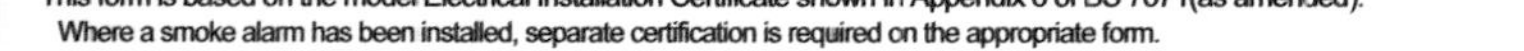

† All boxes must be completed. '✓' indicates that an inspection or a test was carried out and that the result was satisfactory. 'N/A' indicates that an inspection or test was not applicable to the particular installation.

This form is based on the model Electrical Installation Certificate shown in Appendix 6 of BS 7671(as amended).
Where a smoke alarm has been installed, separate certification is required on the appropriate form.

Figure 3.23 Specimen Electrical Installation Certificate: schedule of inspection.

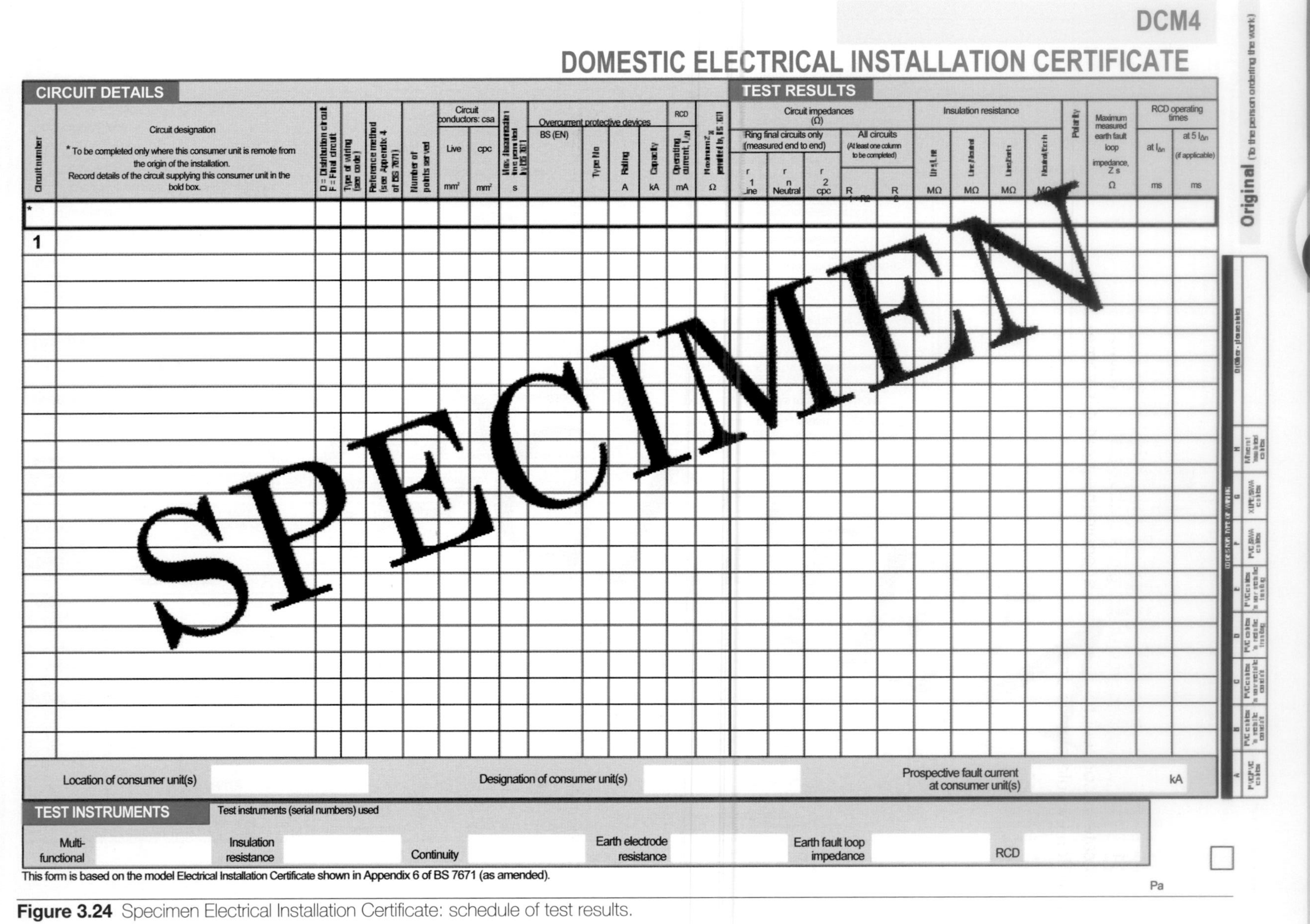

DCM4

Original (To the person ordering the work)

DOMESTIC ELECTRICAL INSTALLATION CERTIFICATE

CIRCUIT DETAILS															TEST RESULTS												
Circuit number	Circuit designation * To be completed only where this consumer unit is remote from the origin of the installation. Record details of the circuit supplying this consumer unit in the bold box.	D = Distribution circuit F = Final circuit	Type of wiring (see code)	Reference method (see Appendix 4 of BS 7671)	Number of points served	Circuit conductors: csa Live mm²	cpc mm²	Max. disconnection time permitted by BS 7671 s	Overcurrent protective devices BS (EN)	Type No	Rating A	Capacity kA	RCD Operating current, $I_{\Delta n}$ mA	Maximum Z_s permitted by BS 7671 Ω	Circuit impedances (Ω) Ring final circuits only (measured end to end) r_1 Line	r_n Neutral	r_2 cpc	All circuits (At least one column to be completed) R_1+R_2	R_2	Insulation resistance Line/Line MΩ	Line/Neutral MΩ	Line/Earth MΩ	Neutral/Earth MΩ	Polarity	Maximum measured earth fault loop impedance, Z_s Ω	RCD operating times at $I_{\Delta n}$ ms	at 5 $I_{\Delta n}$ (if applicable) ms
*																											
1																											

Location of consumer unit(s) | Designation of consumer unit(s) | Prospective fault current at consumer unit(s) kA

TEST INSTRUMENTS — Test instruments (serial numbers) used

Multi-functional | Insulation resistance | Continuity | Earth electrode resistance | Earth fault loop impedance | RCD

This form is based on the model Electrical Installation Certificate shown in Appendix 6 of BS 7671 (as amended).

Pa

Figure 3.24 Specimen Electrical Installation Certificate: schedule of test results.

As we have seen, it is possible, and often preferred, that the microgeneration system is connected to the grid by using a protective device in the consumer's unit. Of course this would require the consumer's unit to have a spare way.

When using this method of connection it would be advisable to carry out a periodic inspection on the existing installation just to ensure that it is safe to connect to; this is not a requirement but it is something which I would always do.

To enable the installation to be connected to the grid and for your customer to benefit from the feed-in tariffs, permission from the distribution network operator (DNO) will need to be obtained. This will require a certificate to be completed as described in the engineering recommendation G83/1. The certificate is confirmation that the small-scale electrical generation (SSEG) installation has been commissioned correctly.

The information required is as follows:

Site details

- Property address and telephone number
- Customer supply number (MPAN)
- Distribution network operator (DNO).

Contact details

- Small scale electrical generation owner
- Contact person and telephone number.

SSEG details

- Manufacturer and model type
- Serial number of SSEG
- Serial number/version of software (where used)
- SSEG rating in amperes and power factor under normal operating conditions
- Maximum short circuit current in amperes
- Type of prime mover and fuel source
- Location of SSEG unit within the installation
- Location of isolation point.

Installer details

- Name and address of installer
- Accreditation / Qualifications
- Contact name
- Telephone number/Email address.

Required information checklist

- Schematic drawings of system
- Small scale electrical generation test report
- Schedule of protection settings
- Make and model of electricity metre.

Installers declaration

This is really a chart with tick boxes which indicate that the necessary requirements have been met.

The small scale electrical generation (SSEG) complies with engineering recommendation G83/1.

All protection settings are compliant with engineering recommendation G83/1.

All protection settings are protected from alteration unless written agreement is obtained between the DNO and the customer.

Safety labels have been fitted in accordance with G83/1.

The installation complies with BS 7671 and all inspection and test certificates have been completed.

Comments on the installation.

CHAPTER 4

Heat pumps

Heat pumps are a very efficient method of heating water and there are two basic types. An air to water heat pump which takes the latent heat in the air and a ground source heat pump which takes the latent heat from the ground.

Air to water heat pump

The basic principle of the operation is that residual heat is taken from the air, usually outside air, although sometimes it is taken from exhaust air such as that from commercial kitchens or air conditioning units. This air is passed through a heat exchanger (evaporator) which is made up sealed coils full of refrigerant which has a very low boiling point. The temperature of the outside air passing over the coils causes the refrigerant to evaporate, the vapour is then passed through a compressor which raises its temperature to a usable level.

This is much the same as pumping up a bicycle tyre; when you compress the air in the bicycle pump, the pump gets quite hot because the air in the pump is being compressed and some of the heat from the air is absorbed by the pump body.

When the vapour has been compressed and its temperature has increased it is passed through another heat exchanger (condenser) where the heat from the vapour is transferred into the water which is being used for heating purposes. As the heat is taken from the refrigerant the vapour cools and returns to liquid, although it is still under pressure. The refrigerant fluid is then allowed to pass through an expansion valve which allows the fluid to expand, which in turn reduces the temperature of the fluid. The cycle is now repeated (Fig. 4.1).

Air source heat pumps can deliver water at a temperature of up to 65°C, although this can depend on the temperature of the outside air. This temperature would be sufficient for domestic water applications although supplementary heating may be required if demand is high or the outside air temperature drops very low. When supplementary

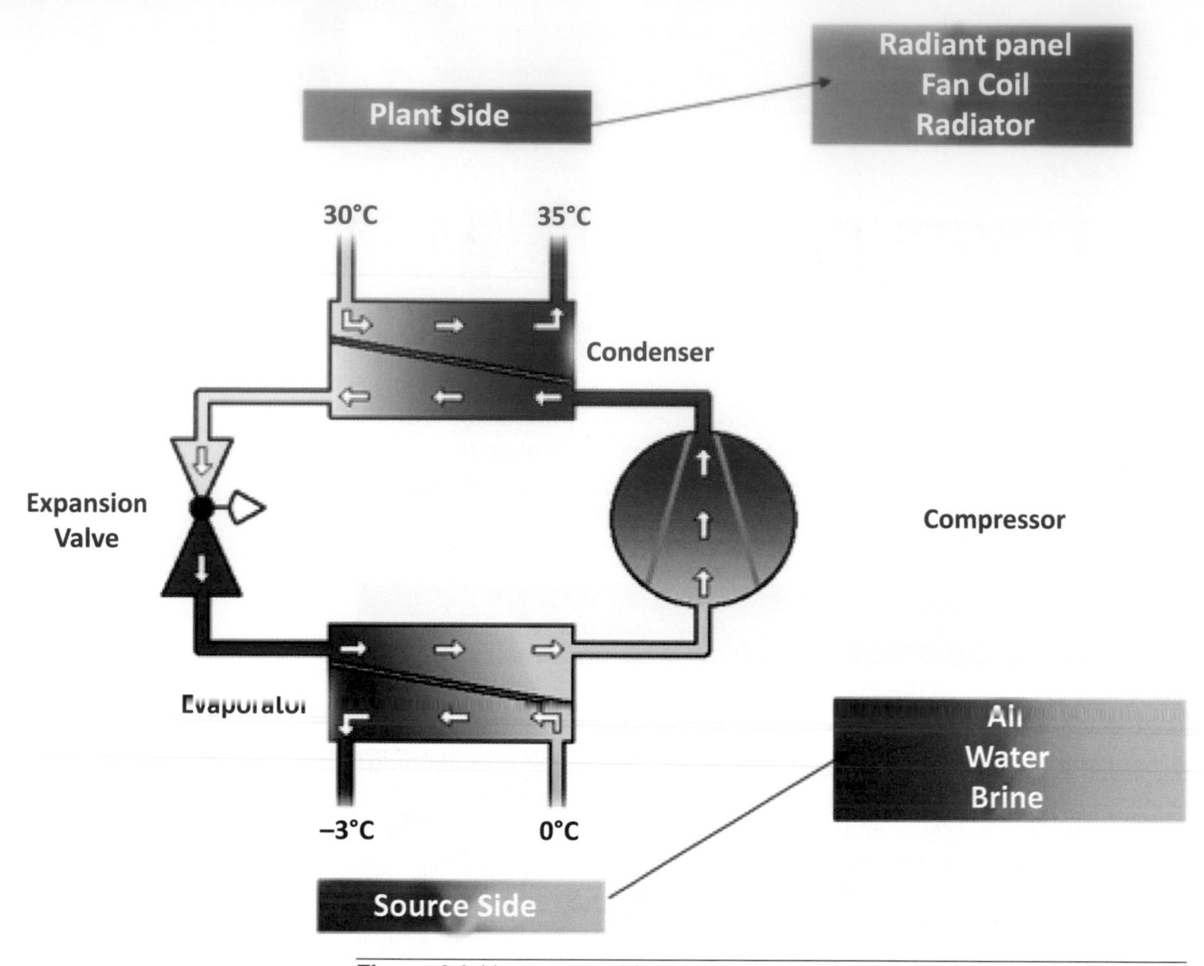

Figure 4.1 Heat pump cycle.

heating is required it is usually provided by an immersion heater. In these situations the efficiency of the system will be significantly reduced, and for that reason it must be carefully monitored.

Air source heat pumps are best suited for use with space heating which utilises under floor heating as this type of system requires a water temperature of around 35°C. As an under floor heating system operates at a lower temperature than that required for domestic hot water, the efficiency of the heat pump is increased. This is because where a low water temperature is required, the unit will not have to work so hard and the efficiency will rise. Of course the use of modern technology will ensure that efficiencies and outputs will increase, probably very quickly, particularly if there is a rise in demand for this type of heating.

Although space heating by this type of heat source is best suited to under floor heating it is also possible to use fan coils which would require a water delivery temperature of 35 to 55°C. It would not be

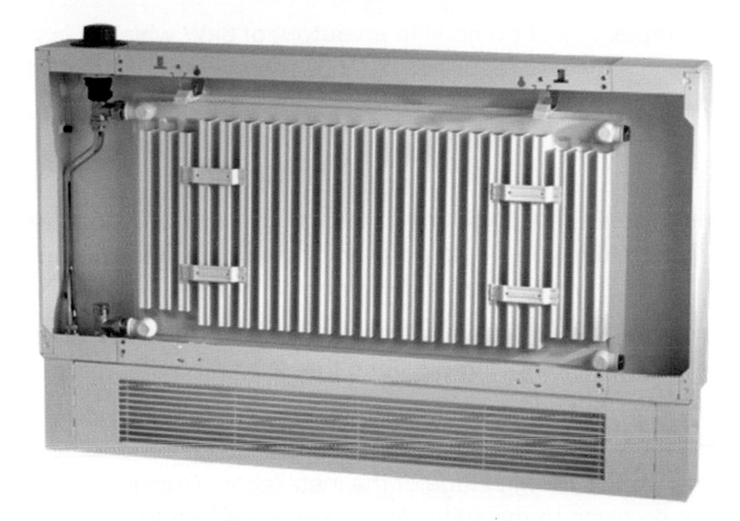

Figure 4.2 Low surface temperature radiator which is designed to operate at 45 to 50°C. This is considerably lower than a standard radiator, which is designed to operate at temperatures around 80°C.
Source: Stelrad Radiators

possible to achieve the temperature required to provide space heating through a conventional radiator system as the radiators would only feel warm, they would not provide enough surface area to properly heat a room. It would of course be possible to increase the size of the radiators but this would not really be economically viable, and the radiators would take up most of the wall space within a room. It is possible to fit low surface temperature radiators which are made specifically for use with heat pumps. These types of radiators are designed to operate with a water delivery temperature of 45 to 55°C. Figure 4.2 shows a low surface temperature radiator with the front cover removed.

Air source heat pumps are incredibly efficient and the efficiency of a heat pump is measured by its coefficient of performance (CoP). This is the heating output in kW divided by the total power consumed by the system (this includes energy required for the motor driving the heat pump, circulating pump and any other controls which may be used).

The European standard used for testing heat pumps is BS EN 14511-2 : 2004 (air conditioners, liquid chilling packages and heat pumps with electrically driven compressors for space heating and cooling). Compliance with this ensures that all heat pump efficiencies are provided using the same process so that the purchaser knows exactly what they are getting for their money.

The calculation for a heat pump CoP is:

$$\text{CoP} = \frac{\text{Heating output}}{\text{Total input power}}$$

As an example, a heat pump with an output of 6kW which is installed into an installation and requires 1.2kW of energy to operate would have a CoP of:

$$\frac{6}{1.2} = 5$$

or put another way, for each unit of energy consumed by the heat pump and controls you would get 5 additional units back giving you a ratio of 1:5. This would mean an efficiency of 500%.

In the example shown we are putting in 1.2kW and getting out 6kW with the added 4.8kW being provided free from air temperature.

There are many advantages to using air to water heat pumps and of course there are also a few disadvantages which need to be considered at the design stage of the installation. Consideration must also be given to the suitability of the property into which the system is to be installed.

There is no doubt that the best investment towards energy saving in any building is insulation. Unless the property is insulated to a high level the efficiency of any type of installation will be greatly reduced. Double glazing is also very important as drafts and heat lost through poorly fitting windows will also reduce the efficiency and comfort levels of any type of installation.

The type and size of the property is very important as, with any other type of heating system, heat losses must be calculated as accurately as possible to allow the installer to fit equipment with a suitable heat output.

Type of building construction, room heights, window types, insulation levels, ventilation and even geographical siting are all things that need to be considered. At present the maximum output for a unit connected to a single phase electrical supply is around 16kW, for outputs greater than that, a three phase supply would be required. 16kW would be sufficient to heat a building with a floor area of up to 300m^2 providing the building is insulated to the levels required by part L of the building regulations. It may be that during very cold days supplementary heating may be required for the hot water requirements. As previously mentioned, an immersion heater could be used and connected to an off peak tariff to reduce the drop in the system efficiency.

Planning permission is not normally required for air source heat pumps although it is always better to check with the local planning authorities before installation begins just to be on the safe side (Fig. 4.3).

Where an air source heat pump is used noise can be a problem. If sited very close to other properties, it is recommended that they are

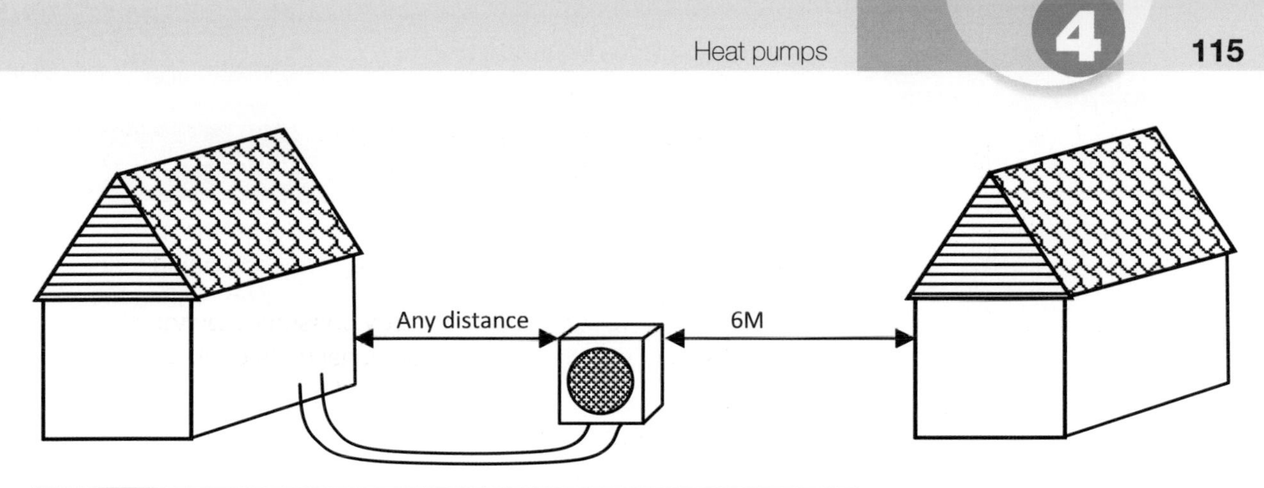

Figure 4.3 Noise has to be taken into account where there are neighbours.

positioned at least 6m from other habitable buildings. Consideration should also be given to the positioning of the unit as problems can be caused by rubbish collecting around the air intake, or items being placed in front of the air intake, which would restrict the air flow through the heat exchanger (Fig. 4.4). In some busy urban areas rubbish would be drawn into the air intake more frequently than perhaps a unit situated in a rural environment.

During very low temperatures problems can arise when ice forms on the outdoor heat exchanger due to the air condensing; this will dramatically reduce the efficiency of the unit although it is unlikely to result in any permanent damage. Most heat pumps are programmed to deal with this situation automatically by reversing the flow of the refrigerant for a short period, this will stop the unit from producing a heat output for a short time.

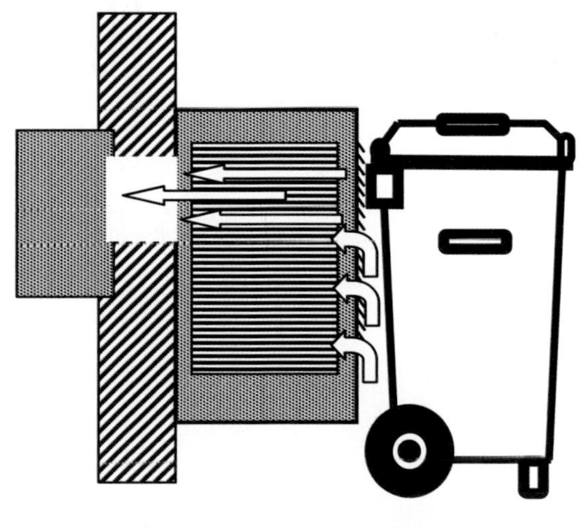

Figure 4.4 Any restriction to the air intake will reduce the efficiency of the system.

Ground source heat pumps (indirect system)

The heat pump used for this type of system works on the same principle as the air to water heat pump, the difference being that the heat is obtained from the ground by a sealed loop ground heat collector which contains a water and antifreeze solution. The heat which is taken from the ground is transferred from the fluid in the ground loop heat collector via a heat exchanger to the refrigerant in the heat pump (Fig. 4.5).

This type of system uses the latent heat in the ground or even a water source such as a lake or an underground water source.

The most common type of ground source heat pump is a closed loop system. This system is one where a long length of tube which is filled with an antifreeze solution is placed in the ground. The antifreeze then absorbs the heat from the ground.

This system consist of three separate elements which together form a very carbon efficient form of space and water heating in energy efficient homes.

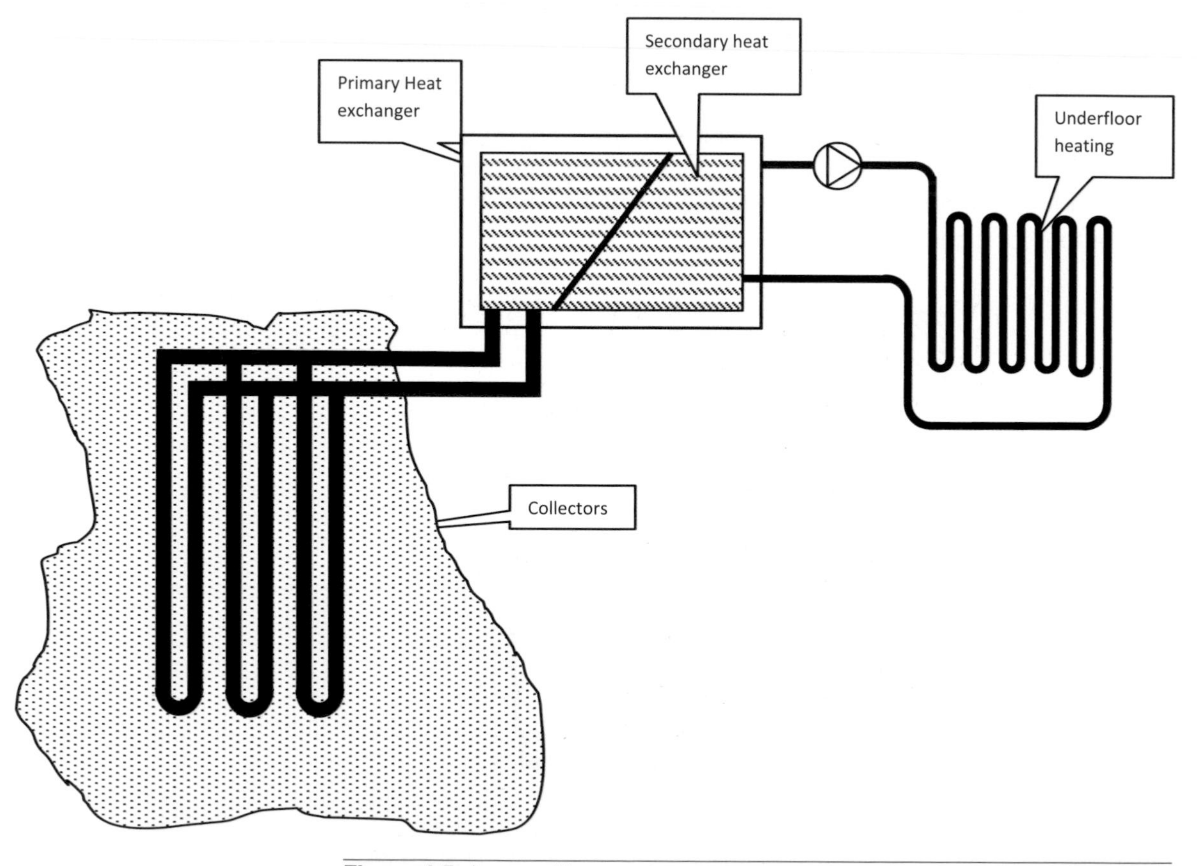

Figure 4.5 A ground heat exchanger.

The elements are (Fig. 4.5):

- a ground heat exchanger
- a heat pump
- a system which will distribute the heat, such as an under floor central heating system.

A heat exchanger is used to extract the residual heat in the ground, and is constructed of a very strong, high density polythene tube. The diameter of the tube is usually dependent on the length of the collector as shown in Table 4.1.

The length of a collector should not exceed 400m; if this length of collector is not enough for the required output then two collectors or more must be installed.

In general the area of land required in m^2 is usually equal to the linear length of the collector, for example a 400m long collector would require $400m^2$ of land, although soil conditions would affect the length of the collector and the area required. A good rule of thumb guide is that the area required for the collector is two and a half times the floor area of the building to which it is going to be connected.

A building with a floor area of $200m^2$ would require a land area of $500m^2$ ($200 \times 2.5 = 500$).

The heat exchanger is filled with a mixture of water and antifreeze called the circulating fluid, as it is important the fluid does not freeze. The solution used is usually a mixture of glycol and water. There are two types of glycol and care has to be taken to ensure that the correct type is used and this of course will depend on the type of installation.

Ethylene glycol can be used but great care must be taken to ensure that it cannot possibly come into contact with any water which is to be used for consumption, in most cases it is far better to use propylene glycol. The characteristics of the different types of glycol is shown in Figure 4.6.

It is very important that the correct ratio of glycol to water is used and this can be tested using a refractometer.

It is common for the mixture to be such that it will prevent freezing to −15°C, but in all cases the freezing point of the circulating fluid

Table 4.1

Length of the collector	Overall diameter of collector
100 metres	25mm
250 metres	32mm
400 metres	40mm

Property of types of Glycol	Propylene Glycol	Ethylene Glycol	Points to consider
Freezing point	Better	Good	Ethylene glycol can't carry as much heat as propylene glycol. More fluid must be circulated to transfer the same amount of energy. Pumps volume increased.
Efficiency of heat transfer	Good	Less	Ethylene glycol can't carry as much heat as propylene glycol. More fluid must be circulated to transfer the same amount of energy. Pumps volume increased.
Viscosity	Higher	Lower	
Flammability	Low	Low	
Biodegradability	Will degrade within a maximum of 30 days	Will take longer than 30 days to degrade	Propylene glycol is better for the environment
Toxic	Low level	Acute when swallowed	Ethylene glycol must never be used in any drinking water or food processing system
Carcinogenic	No	No	A carcinogen is any substance or agent which promotes cancer
Skin irritant	Low	Low	Propylene glycol is used in small amounts in cosmetics
Effectiveness	Good	Better	More antifreeze is needed of propylene glycol to achieve the same freeze point

Figure 4.6 A table showing the characteristics of different types of glycol.

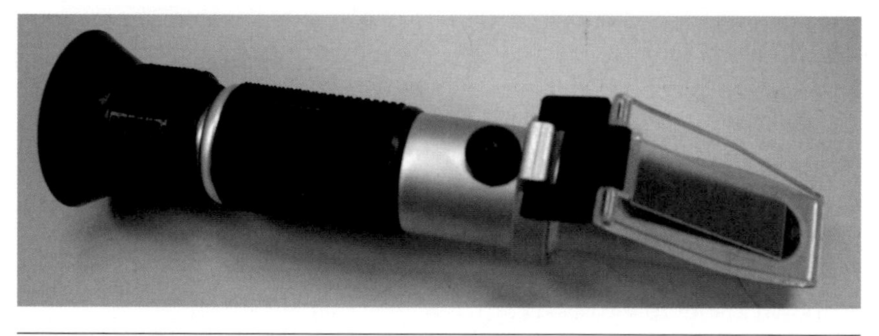

Figure 4.7 A refractometer can be used to test that the system has the correct ratio of glycol to water.

should be at least −5°C below the average of the inlet temperature and the outlet temperature of the source side of the heat pump.

Manufacturer's instructions should be followed with regard to the fluid used as, if the viscosity of the fluid increases, more energy will be required to circulate the fluid around the heat exchanger. This of course will reduce the efficiency of the system.

Heat exchangers can be installed horizontally in the ground using a trench system or they can be installed vertically using a borehole system.

Horizontal trench system

The horizontal trench system is very common as it is the easiest to install, although the disadvantage of this type of system is that it requires a large area of land for the installation of the pipework forming the collector. Clearly the more collector that is in contact with the soil the greater the output of the system will be. Figure 4.8 shows a large excavated area with the collector in place and covered with a layer of sand before the excavation is backfilled.

Problems can occur when an installation is too small and trying to take more heat from the ground than it has been designed for. If the ground around the pipe has too much heat taken from it, the soil will freeze. As we know when anything freezes it expands, and contracts again when it thaws out. In these cases the soil would shrink away from the heat collector and this of course would result in the surface area of the collector which is in contact with the soil being reduced (Fig. 4.9a/b). This in turn would affect the efficiency of the system and reduce it considerably.

Before installing a collector in the ground the length required for the correct operation must be calculated, but this calculation will normally be carried out by the manufacturer of the equipment being used. Heat pump design is based on keeping a building at a constant

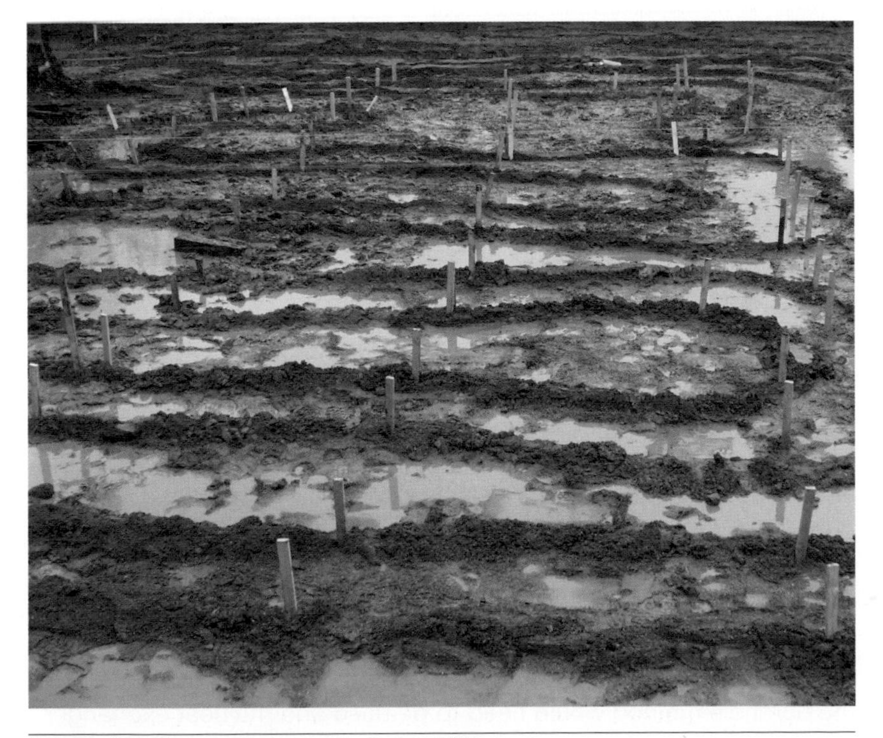

Figure 4.8 Pipes in place, ready for backfilling.

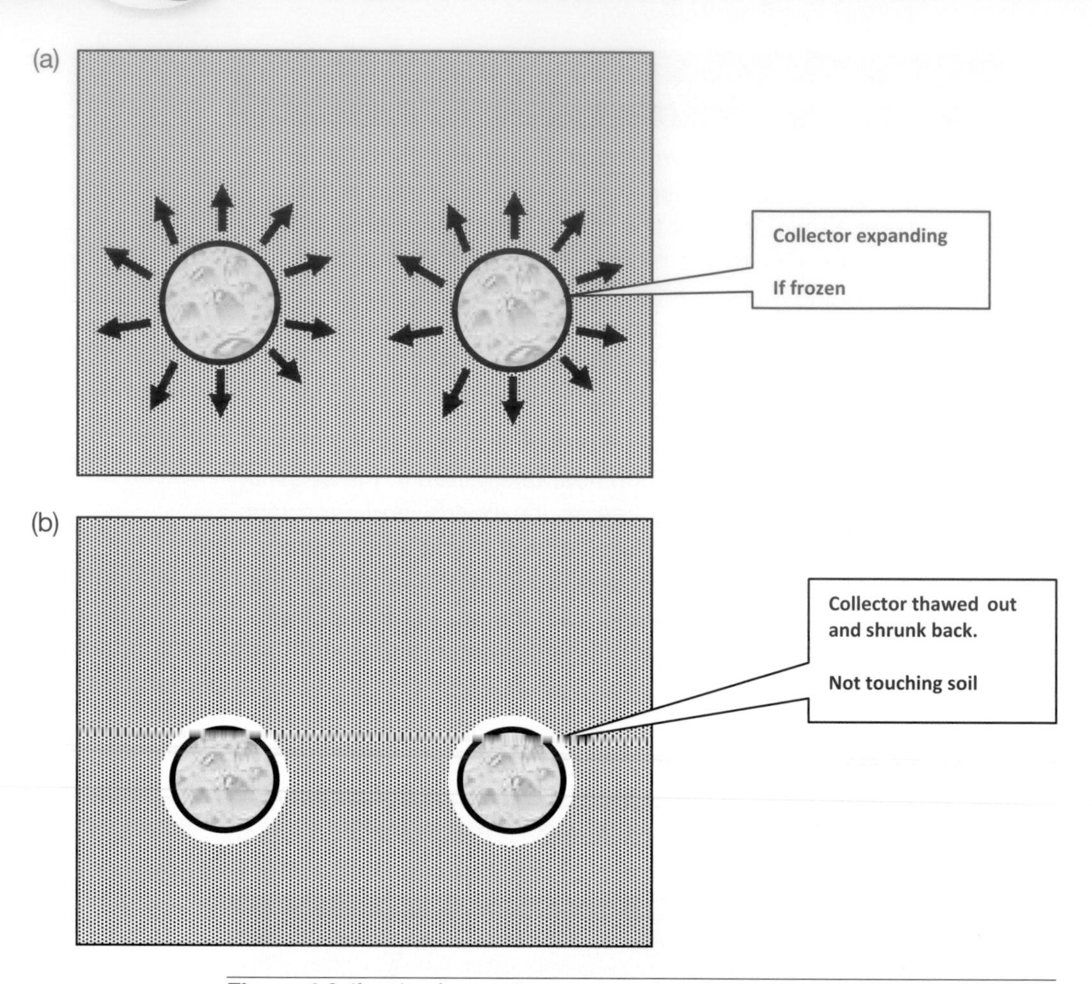

Figure 4.9 If a pipe freezes it expands and pushes the soil away. When it contracts the pipe will shrink and the soil will probably not shrink back with it.

temperature so the heat pump is sized taking into consideration the heat losses from the building, the temperature required for the heating and any hot water requirement.

Consideration must be given to where the collector is to be sited. You must remember that the collector is going to take the heat out of the ground so for the system to operate properly the siting must be in the open to allow the sun to replenish the heat taken from the ground. Any collectors which are installed under tarmac or in the shadow of trees or buildings for large parts of the day will not be very efficient (Fig. 4.10).

The horizontal method will require a depth of between 1.5 and 2 metres, but if there are large boulders or areas of bedrock within 1.5 metres of the surface of the proposed site the use this method will not be possible if good results are to be expected. In these situations the borehole method would need to be used and the heat exchanger installed vertically through the rock.

Figure 4.10 The ground collects its heat from the sun so a shadow cast over the ground will reduce the amount of heat collected which in turn will affect the efficiency of the heat pump.

The temperature of the soil at the depth at which the heat exchanger is to be installed will need to be measured, as of course this is important when calculating the length of heat exchanger that would be required for a specific heat output. As a general rule the temperature of the soil up to a depth of 2 metres will change depending on the seasonal air temperature with a time delay of about one month. This of course means that the soil temperature will be at its highest temperature when it is not required, and at around its lowest temperature when good transference of heat from the soil to the heat exchanger is required. Providing the system is calculated and installed correctly this will not present a problem. Once a depth of around 10 metres is reached the ground temperature remains pretty constant at around 10°C to 14°C, this will depend on local conditions.

The type of soil will also have an effect on the size of the heat exchanger. The best type of soil would be wet and compact, as this would have a high thermal conductivity, whereas loose dry soil may have air pockets and would have a very low thermal conductivity.

An electric pump is used to circulate the fluid around the pipework and this pump should be sufficient to cause a good turbulent flow. An important factor in this is that the pipework must be the correct diameter to suit the size of the system and the pump sized accordingly.

If the pipe diameter is too small the pump will have to work hard, which in turn will result in it using more energy, which defeats the object, as the whole idea is to be as energy efficient as possible.

The diameter of a collector pipe for horizontal systems is usually between 20mm and 40mm and the pipework for the heat exchanger can be laid in the ground, often with a number of them in parallel side by side in a single trench. It is very important that the pipes are not too close to each other as the soil either side of them will need to be able to conduct heat into each pipe, to ensure the maximum heat from the soil is collected. As a general rule the pipes should be a minimum of 300mm apart, preferably more where possible (Fig. 4.11).

A simple calculation will show that the trench will need to be a minimum of 1.5m wide.

Six pipes with a space of 300mm between them will require a trench with a minimum diameter of 300mm × 5 = 1500mm or 1.5 metres.

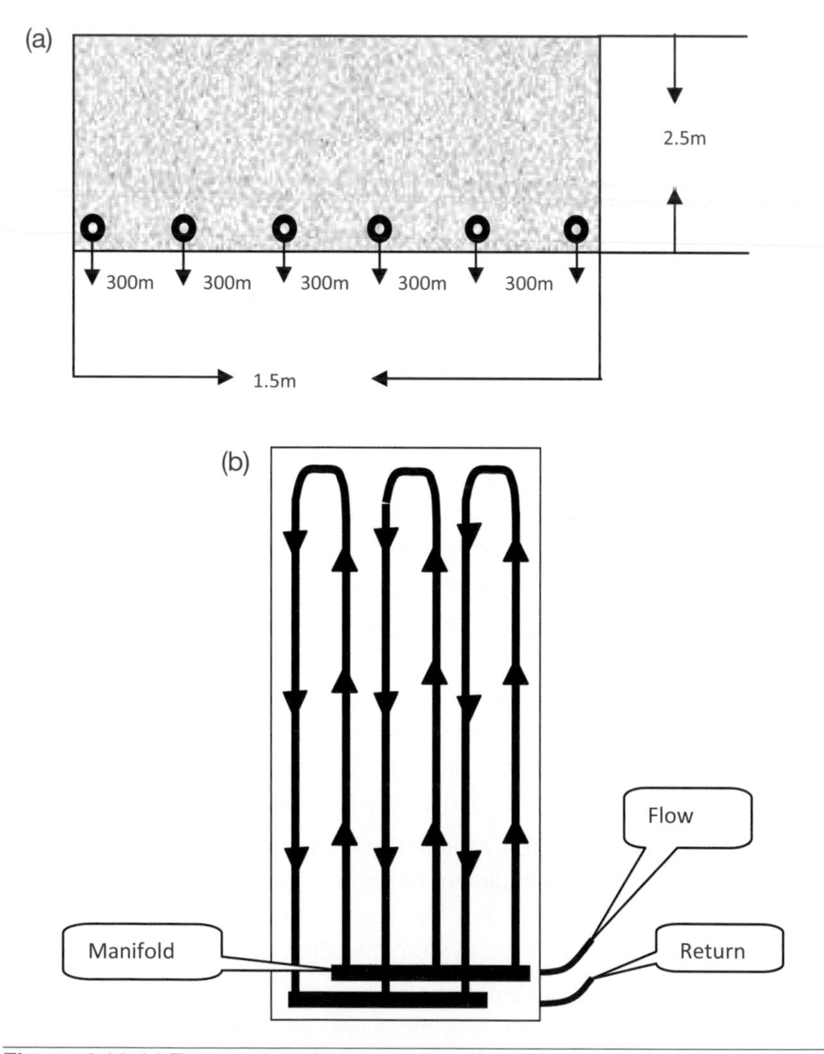

Figure 4.11 (a) The spacing of collector pipes for correct operation of the system. (b) Collector pipes connected to a manifold often provide good efficiency.

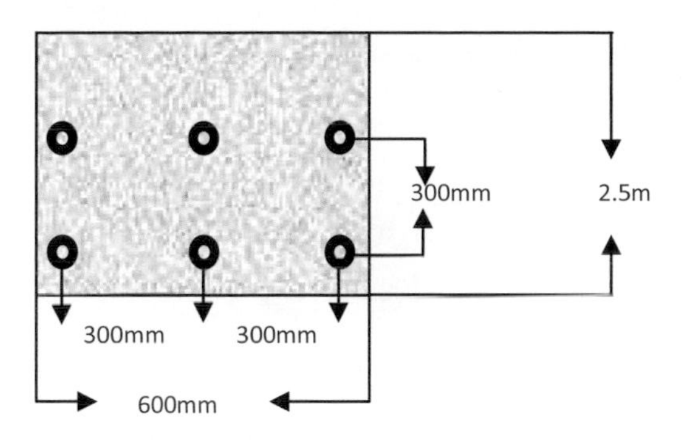

Figure 4.12. If the trench is deep enough pipes can be installed in layers.

It is possible to install the pipes on top of one another providing the trench is deeper, this of course would not require such a wide trench (Fig. 4.12).

Where space is limited another method of installation would be to use a pipe as a series of over lapping coils, this is sometimes referred to as a *slinky* (Fig. 4.13a). Where this method is used it is possible to lay the pipework flat along the bottom of a trench, or where a narrower trench would be more suitable the pipework can be installed vertically (Fig. 4.13b).

Where a collector is to be laid in the ground it is important that the bottom of the trench has a 150mm layer of sand in it before the collector is placed on it, ideally the collector is then covered with

Figure 4.13 (a) A slinky is a coil of pipe used as a collector.
Source: Caron Alternative Energy

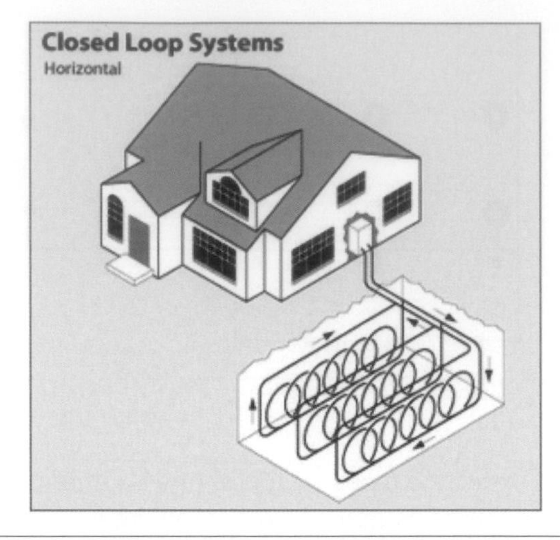

Figure 4.13 (b) A slinky coil can be installed vertically if required.

another 150mm layer of sand before backfilling the trench. When backfilling, the first 300mm of soil should be placed by hand, and be free from any sharp stones which may over a period of time damage the collector (Fig. 4.14). Any soil/backfill above this may be mechanically placed and compacted.

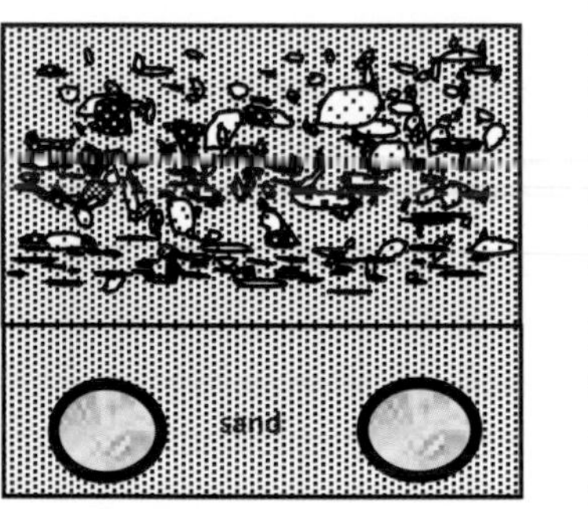

Figure 4.14 It is important that the first 300mm of backfill material is placed into the trench very carefully.

Where the collector passes over or under a drain or water pipe it must be insulated for a distance of 1.5m either side, this is to prevent the drain or water pipe freezing. The collector must also be insulated for a distance of 1.5m from the entry into the heat pump, failing to do this could result in the ground freezing and moving. Great care must also be taken when bending the pipe and a 40mm pipe should not be bent with a radius of less than 1m (Fig. 4.15).

As this type of installation will require excavation, all aspects of health and safety must be considered and all regulations must be

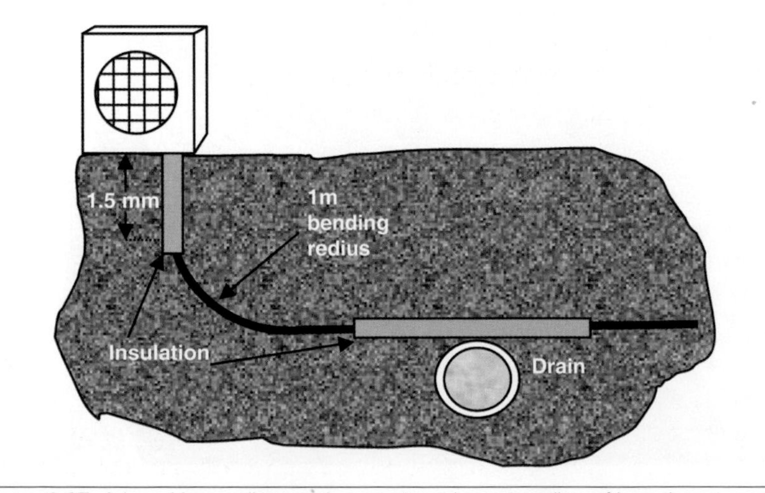

Figure 4.15 A bend in a collector pipe must not have a radius of less than 1 metre.

complied with. A risk assessment should always be carried out before commencing any type of work and this of course will highlight any risks and allow precautions to be put in place to prevent accidents.

Working in any kind of excavation can be a dangerous exercise, so the confined spaces regulations 2005 require that a trench deeper than 1.25m must be shuttered to prevent the collapse of the walls of the trench.

To ensure that work is carried out safely it is necessary to think very carefully and document the risks involved in any work which is to be carried out, the site conditions and the type of tools which are going to be used. This is called a risk assessment.

There are five steps to a risk assessment.

Step1. Identify the hazards

A hazard is anything which has a potential to cause harm.

This may involve asking people around you what they think could be considered a risk.

Manufacturers' literature should be consulted as this will help identify any hazards which may be present due to the type of equipment being used. Control of substances hazardous to health (COSHH) data may also be provided by manufacturers for materials used in the installation of the products used.

Not only immediate hazards should be considered, anything which could cause long term effects must also be assessed. These could be noise or vibration along with many others.

Hazards to be considered when excavating would be.

- Noise from digging equipment.
- Fumes from digging equipment or even the soil.
- Injury from the movement of machinery.
- Lifting/moving heavy equipment.
- Falling into the trench (people or equipment).
- The trench collapsing on to whoever is working in it.
- Damage to other underground services such as electricity, gas or water.
- Is there a possibility that any excavation may cause movement in any structures nearby?

Step 2. Decide who might be harmed

Most places of work have many people who are involved in different activities at different times, some workers may be very experienced while others may be very young with little or no experience at all. Consideration must also be given to people with disabilities. It is important to consider what is going on and who may be harmed.

- Ask people who are working on site who they think may be at risk.
- Is the area accessible by the public, if so could they be hurt by anything which is under your control?
- Are there any workers, visitors who may have access to the site who may not be there all of the time and may be unaware of what is happening?
- Are there any workers or people with disabilities who have access to the site and may be at a particular risk?

Once the hazards and those who are at risk have been identified go to step 3.

Step 3. Evaluate the risks and decide on the precautions

The law requires you to do everything you possibly can to protect people from harm.

- Can the risk be eliminated altogether?
- Can the risk be controlled so that harm is unlikely?

The Health and Safety Executive produce a document which helps us to decide on the best way to eliminate or control risks. The document suggests that we apply the following principles, preferably in the order as shown.

- Try a less risky option.
- Prevent access to the hazard by barriers.
- Organise work to reduce the exposure to the hazard (place barriers between those who need to work in an area and those who do not).
- Ensure the correct PPE is worn.
- Provide welfare facilities, such as provision for washing off any contamination.
- Provide first aid facilities.

Step 4. Record your findings and implement them

When carrying out a risk assessment it is important that your findings are documented, this is because circumstances and places of work change. The risks can then be reviewed and updated as required. Where an organisation employs more than five members of staff it is a legal requirement that risk assessments are kept in written form.

It is not necessary to write reams and reams of information, the best risk assessments are the simple ones. They must show that:

- a proper check was made
- you asked who might be affected
- you dealt with the hazards
- the precautions are reasonable and that any remaining risk is low.

On completion of a risk assessment it is usual to make a plan of action which will enable you to deal with any risks in a sensible order. A plan of action would include:

- A simple list of improvements which could be carried out quickly and cheaply to provide a temporary solution until long term permanent measures could be put into place.
- Long term solutions to any risks which are most likely to cause injury or ill health.
- Areas where employee training would reduce the risks identified.
- A review form with dates and names.

Step 5. Review your risk assessment and update if necessary

Most work places are subject to change due to various circumstances, for example the introduction of computers into the work place. Even something as simple as this has introduced new hazards which have to be addressed.

Due to changes, even very small ones, risk assessments must be continually reviewed. It is recommended that risk assessments are reviewed and updated at least annually, or in the event of an accident or even a close shave they should be reviewed more often.

Lake and canal collectors

As explained earlier in this book, the wetter the soil the more conductive it is, water is an even better conductor! Where there is a lake close to a property, heat can be collected from it by using a closed loop system placed on the bottom of the lake and delivered to a heat pump which is linked to the heating system within a building. Figure 4.16 shows a typical closed loop collector during construction.

Figure 4.16 Closed loop collector during construction.

Closed loop system

Any lake used would need to be large enough to collect enough solar energy to provide sufficient heat energy for the installation, the size of the collector would normally be designed by the suppliers of the heat pump, and of course if the lake was not large enough then it may not be able to sustain the required water temperature, particularly during very cold periods. Where the collector runs from the lake to the house it must be installed using the same method as the horizontal trench system. The lake must also be deep enough not to freeze on the bottom.

A major advantage of this type of system is that it is easy and cheaper to install than other heat pump arrangements. The collector is floated out on to a lake, probably using a small dinghy, and when it is in position it is filled with fluid, which of course makes it sink to the bottom of the lake. Once in place it is held there by weights and over a very short period of time the collector will be covered with silt from the lake and become invisible. Where the system is used in a canal then it is simply a matter of placing the collector into the canal and securing it into position. Figure 4.17a/b shows a completed collector before it is positioned into a canal.

Testing a horizontal closed loop collector

All closed loop collectors must be thoroughly pressure tested before they are covered, this is to ensure that there are no leaks which will cause problems later. They must be filled with water and then pressure tested to 10 bar for a period of 24 hours. Providing there are no leaks the collector can then be placed into position and covered. A pressure test pump which is used for testing central heating systems can be used to test the collector. This will involve sealing one end of the collector and fitting a connector suitable for the pressure test pump to be attached to (Fig. 4.18).

Open loop heat pump

This type of system draws water from a borehole or other water source, the water then passes through a heat exchanger and then back into the ground via a separate borehole (Fig. 4.19). The average temperature of ground water in the UK is 8–10°C and it remains fairly constant throughout the year; this is one very good reason why the system could be an efficient one. One disadvantage is that the water extraction system would require a fairly large pump, which requires a lot of energy to run it, and this of course would reduce the system's efficiency.

Figure 4.17 (a/b) A canal collector laid out ready for installation.

Figure 4.18 All collectors must be pressure tested before they are covered, this is carried out using a pressure test pump.

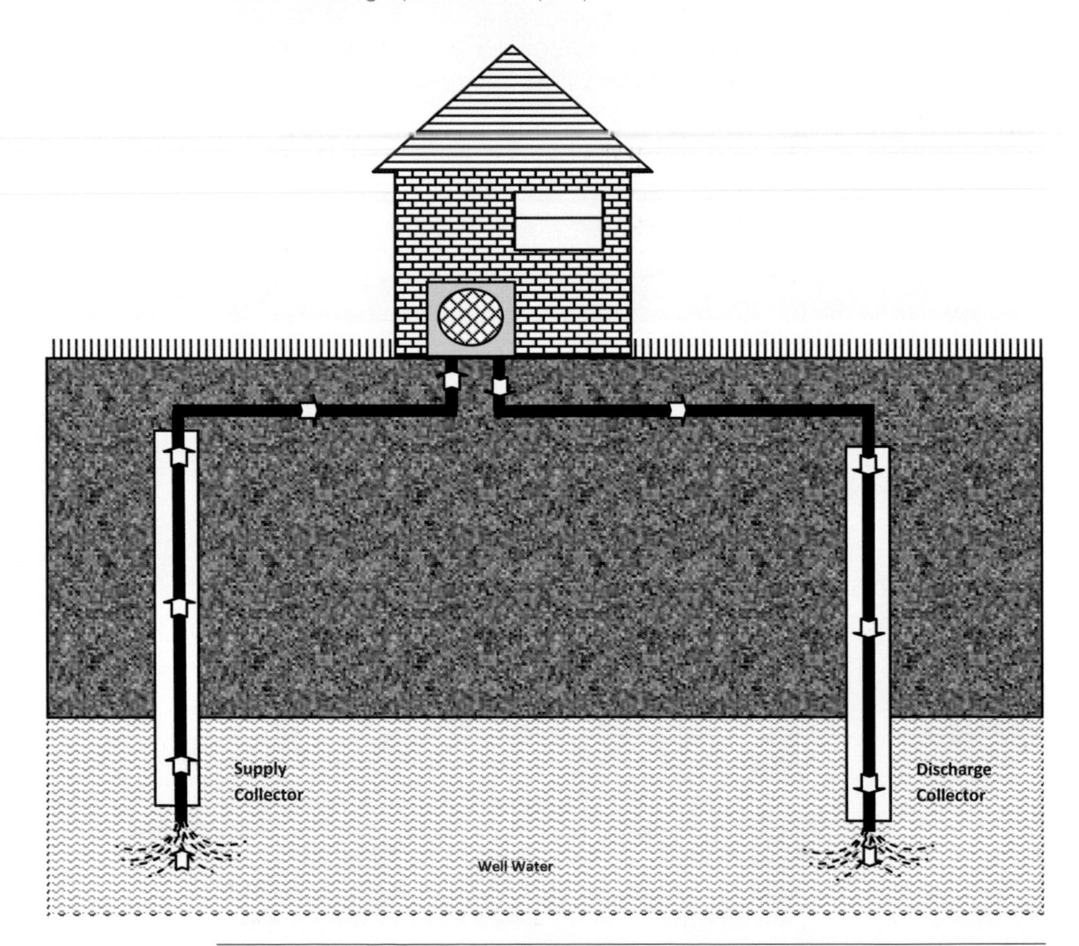

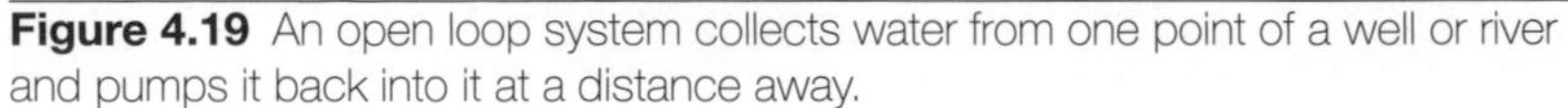

Figure 4.19 An open loop system collects water from one point of a well or river and pumps it back into it at a distance away.

Due to the open loop system using a constant flow of water drawn directly from its source, the heat pump used would normally be designed specifically for this purpose and would have a corrosion resistant heat exchanger inside it. This would allow the water to be pumped directly through it.

Where it is intended to use this type of system the permission of the environment agency must be sought as an extraction licence would be required. Open loop systems are specialist installations which would require a survey to be carried out to confirm that the correct hydrological conditions exist.

An underground water source is known as an aquifer, which is often under pressure. The consequences of drilling into one could be quite extreme and very costly, so this is another very good reason why this type of installation should be carried out by specialists.

Due to the water being drawn from the ground or other water source such as a river, it is necessary to have very good quality filter system in place to remove any grit and other debris from it, before it reaches the pump and causes premature wear on the impellor and any other moving parts which it might come into contact with. Blockages or restrictions to the water flow would reduce the efficiency of the system considerably and the filter system would need to be very accessible to allow for regular cleaning.

Borehole method

Where space is very limited the borehole method would be suitable as it takes up very little space. It is also suitable where there is a layer of bedrock or unsuitable ground conditions which would prevent the use of the trench method.

This method can work out very expensive as a hole has to be drilled vertically through the soil to a depth of at least 15m and sometimes 150m, this of course requires the use of specialist drilling equipment (Fig. 4.20).

The principle is the same as the trench method although better results are usually achieved as the ground temperature is constant at between 10°C and 14°C at this depth.

Where it is not possible to obtain enough energy by using one borehole, two or more can be used and linked together. The distance between vertical boreholes should be a minimum of 6m for a domestic installation. As with any type of ground source system a site survey must be carried out by a specialist before commencing work. The company supplying the heat pump equipment will calculate the length of collector required and this information, along with the

Figure 4.20 Special machinery is used to bore down through the earth to the required depth.

information about ground conditions, would be used to decide on how many boreholes are required.

The boreholes can be drilled vertically or if required at an angle of around 45°. Using a system which has boreholes at an angle is a major advantage where more than one borehole is required. A number of boreholes could be drilled using one point of entry and this is very useful where drilling space is limited (Fig. 4.21).

The collector which is to be inserted into the borehole is usually a 40mm tube formed into a U shape with a wall thickness of 3.7mm. The pipe material and wall thickness is important as when it is inserted into the borehole it will have to withstand pressures of 1 bar for each 10m depth.

Due to the diameter and material from which the tube is made it is not possible to bend the tube into a U shape. A purpose made U bend is attached to the end of two pipes to form a collector of the required length. The U bend must be attached using a method which will withstand the pressure and will also prevent it from leaking.

Remember, once inserted into the borehole the collector will be under pressure for the whole of its working life, which will be in excess of 30 years. The heat pump may well require replacing as it has moving

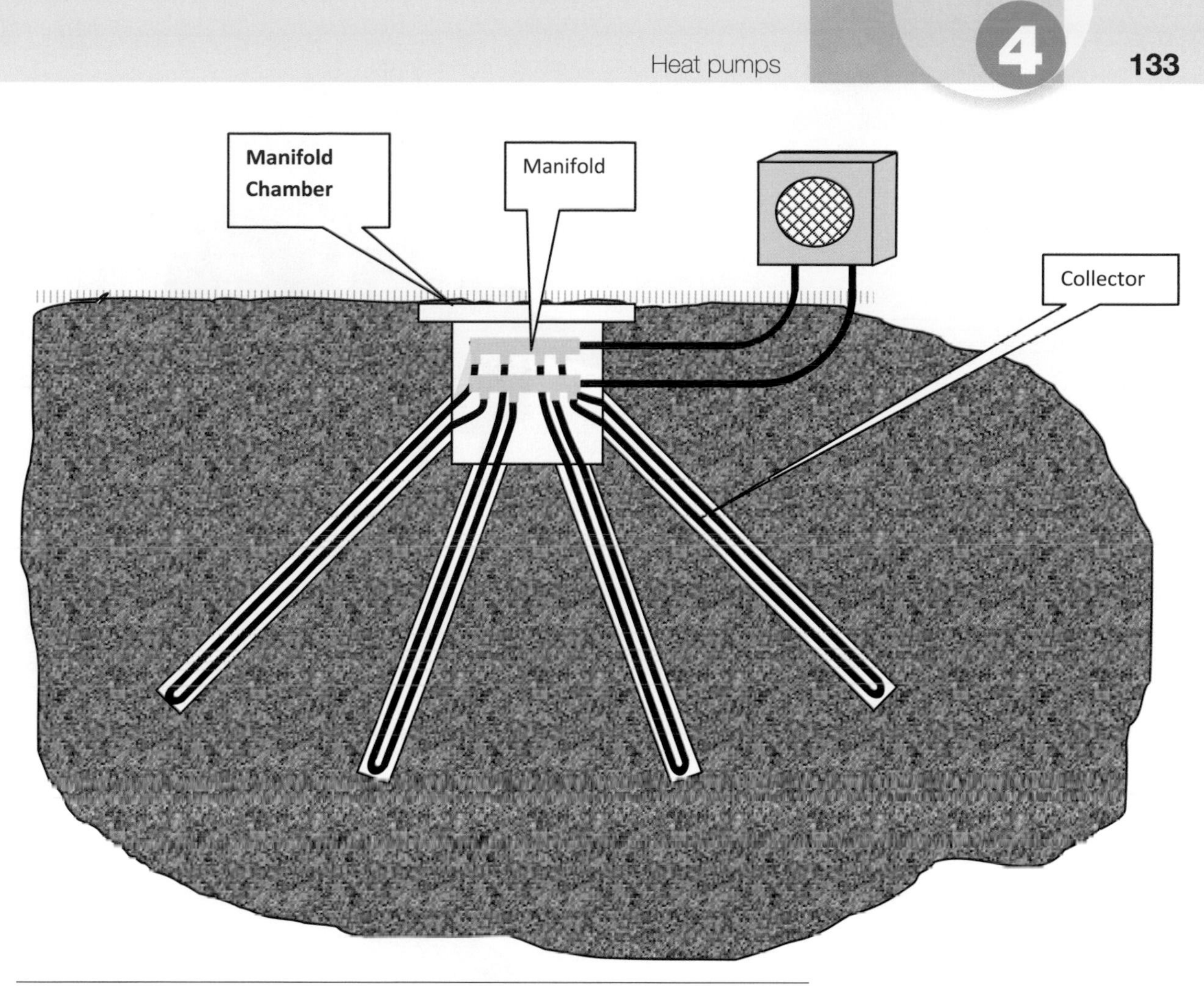

Figure 4.21 One point of entry can be used where space is limited.

parts, but the collector should remain operational for as long as it is required so careful installation will help to make this possible.

The method used to join the U bend to the pipe is known as fusion welding and this can be carried out using a female socket or simply connecting the pipe end to end.

The socket fusion weld system uses a socket with a heating element inserted into it, possibly a piece of copper wire. The end of the pipe is inserted into the socket and a current is passed through the heating element. This heats up and melts the surface of the socket and the pipe. When the material reaches a sufficient temperature it fuses together and forms a very strong indestructible joint.

Another method is to use a tool which heats the outer surface of the pipe and the inner surface of the socket at the same time (Fig. 4.22a/b). After a set time the pipe and socket are pushed together and held steady until the plastic fuses together.

Figure 4.22 (a) Fusion welding using a socket and pipe heating tool.

Figure 4.22 (b) Socket and pipe heating tool used for fusion welding.

Where it is required to join a pipe end to end, a flat plate method is used. The ends of the pipe are cut so that they are clean and square, they are then placed end to end in a tool which holds them secure with a flat plate separating them (Fig. 4.23a/b).

The plate is then heated to a temperature sufficient to melt the pipe material. When the material has reached the correct temperature, the plate is withdrawn and the ends of the pipe are pushed together and held firm, the plastic then fuses together. When the pipe has cooled down it is released from the device and a very strong joint has been made.

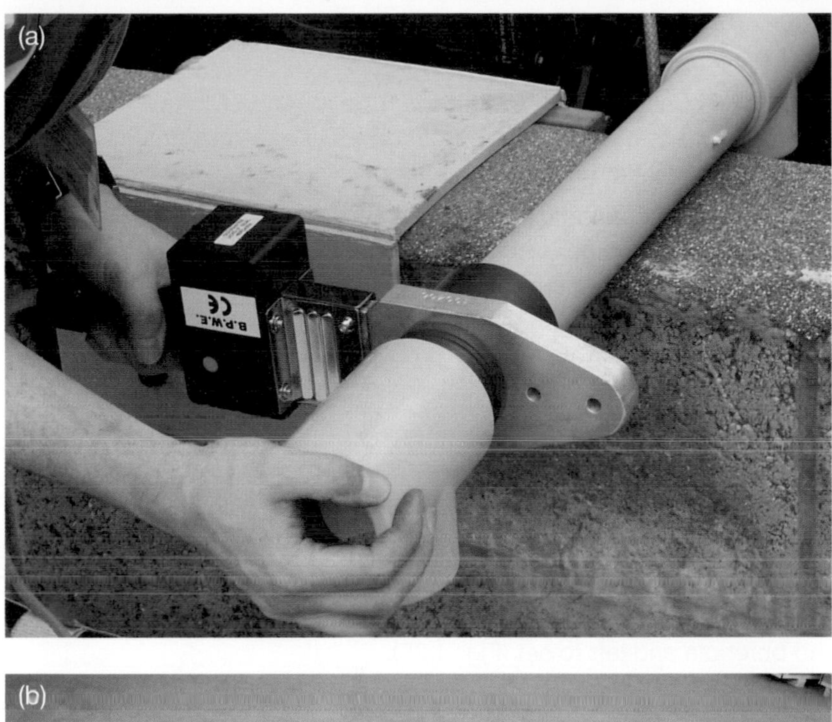

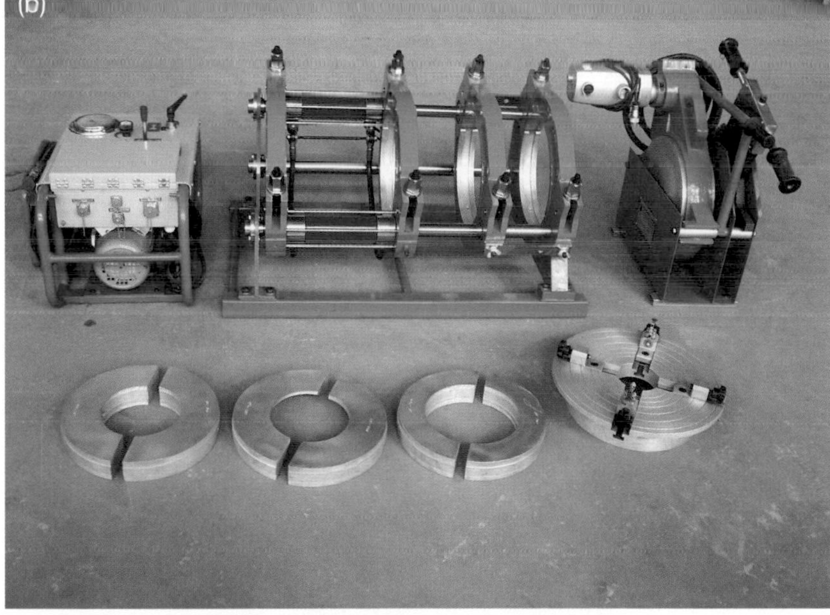

Figure 4.23 (a/b) Flat plate pipe joining tools.
Source: Widos Technology

Installing the collector into the borehole

It is important to test the pipework before it is installed into the borehole and again after installation. It would be a huge waste of finances if the pipe was found to have a leak, for whatever reason, after the job was completed.

Once the U bend has been fitted the tube must be filled with water and pressure tested to 10 bar for 24 hours. If after this time is has no leaks it can be inserted into the borehole.

To carry out this operation it is recommended that a weight of 5kg for every 100m of pipe is attached to the U bend to ensure that the collector drops down to the bottom of the borehole as easily as possible. For example, if the borehole was 75m deep a 150m of collector would be required. This would require a weight of 7.5kg.

Once the collector is inserted into the hole, the hole will need to be backfilled. This will require a grout being poured into the hole which will allow transference of heat from the soil to the collector, the most commonly used material for this is Bentonite. This material is available from any stockist who deals in geothermal products and is delivered in bags in powder form (Fig. 4.24).

It has to be mixed with water to form a grout and then pumped into the borehole and left to set (Fig. 4.25).

The setting time of this material is dependent on the manufacturer: some products have a working time of 30 minutes or so and then

Figure 4.24 Bentonite usually comes in bags similar to cement bags.
Source: Barnes Plastic Welding Equipment

Figure 4.25 A special pump is used to mix and pump the Bentonite into the borehole around the collector.
Source: Special Plasters

set slowly, others flash set after a period of time. It is normally better to use a material which sets slowly as it is far easier to clean any equipment which has been into contact with it. Most suppliers will be able to calculate the amount of grout which is required as long as they are provided with the following information:

- depth of bore
- diameter of bore
- number of bores
- diameter of collector pipe.

When the collector pipework has been backfilled and connected to the heat pump it must be flushed out with clean water. It is important that the antifreeze solution which is to be used to fill the collector is not mixed with the water used for flushing. Although care must be taken to avoid it this water may be contaminated with all sorts of foreign matter which has found its way into the collector during installation. After thoroughly flushing the collector it must be filled with an antifreeze solution.

To fill the collector loop it is usual to have filling station which has a large vessel or tank which is filled with an antifreeze solution (Fig. 4.26). A pipe is then connected to the inlet of a pump with the other end of the pipe placed in the vessel full of solution. A pipe is then connected between the pump outlet and one end of the filling loop,

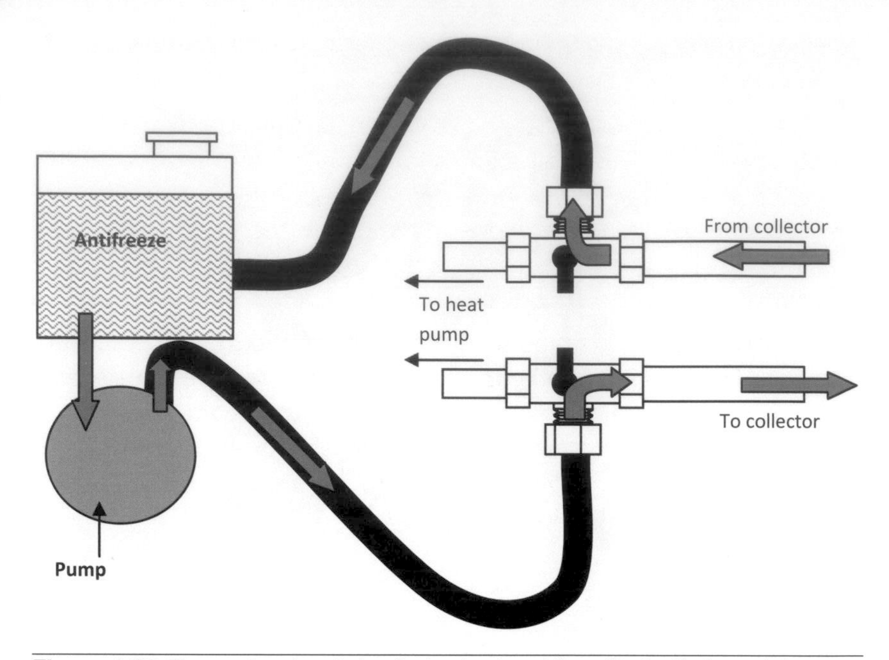

Figure 4.26 The system has to be flushed out and then filled with the solution which is to remain in the collector pipe, this is usually a mixture of glycol and water. Filling and flushing is carried out using the same type of filling station as would be used for filling a solar thermal system.
Source: colcrete-eurodrill.com

another pipe is then connected to the other end of the collector and the pump allowed to run (Fig. 4.26).

The antifreeze solution will then be forced into the collector and the water which has been used for flushing will be pushed out of the pipe which has been connected to the other end of the collector, this pipe should be held over a drain until the antifreeze starts to discharge from it (the vessel containing the antifreeze solution may need to be topped up as the water is being pumped out of the collector). As soon as the antifreeze appears the pipe should be placed into the antifreeze contained in the vessel. The antifreeze should be allowed to pump around the collector for a period of at least an hour although longer would be better, if possible, as this will ensure that all of the air is purged from the system. This flushing is usually carried out after the heat pump has been connected and the filling pump pipework would be connected to filling points in the system pipework.

Direct circulation system

This type of system has a copper ground heat exchanger which is filled with a refrigerant, which eliminates the need for a heat exchanger between the collector and the refrigerant in the heat pump.

This system is known as a direct expansion system (DX system) and is more efficient than indirect systems.

By eliminating the intermediate heat exchanger, the refrigerant's temperature is closer to the ground's temperature, which lowers the heat pump's required compression ratio, reducing its size and energy consumption. Also a shorter ground loop can be used because copper tubing is more efficient at transferring heat than the polyethylene pipe used in conventional closed loops.

The disadvantage of this type of system is that if the system contains more than 3kg of refrigerant it can only be worked on by a registered engineer and must be checked regularly for leaks. It is also possible for the ground around the collector to freeze when the system is calling for a lot of heat. Environmental problems could also be an issue if the collector were to develop a leak.

Commissioning the system

Before the commissioning of the system can begin it is very important to ensure that the installation of the collector and the heat pump is complete.

The heating and domestic hot water system must have been flushed thoroughly then filled and pressure tested. A check must be carried out to ensure that all other equipment which is connected to the system, including any electrical accessories, is installed to comply with the manufacturer's instructions.

Once it has been confirmed that the system is complete and safe to be energised the commissioning process can be carried out. This is usually completed by the manufacturer of the pump or a qualified engineer.

After the heat pump has been commissioned and any documentation which is required by the manufacturer or the renewable heat incentive scheme has been completed, the installer must be provided with information on how to use the system and how to isolate it if required.

CHAPTER 5

Combined heat and power units (CHPs)

This cannot really be considered as renewable microgeneration and it is better described as an energy efficient method of providing heat and power. A correctly installed CHP unit can save up to 30% of a domestic fuel bill.

This technology can be used to provide energy on a small or large scale, but in this book we are only looking at small scale or micro-generation. The principle of the system is that it generates heat and power very close to its point of use, which of course eliminates any losses due to having to transport energy long distances from its source.

One big advantage of a domestic CHP is that the unit would be a direct replacement for a conventional domestic boiler with servicing the same as a conventional boiler.

A combined heat and power unit can use all sorts of different fuels, for this reason they are described as being fuel neutral. This technology is evolving quite quickly as are most other kinds of energy efficient technologies.

Internal combustion engine

One method requires the use of an engine which drives a generator to produce electricity. The engine is normally an internal combustion engine which is used to drive a generator. A big advantage of using internal combustion engines is that they can be adapted to run on pretty much any kind of fuel which will allow them to work efficiently. For obvious reasons biofuels are the preferred choice as they are generally more carbon friendly than most fossil fuels.

As the engine is driving the generator it creates heat and this heat is transferred into a coolant as it passes through the engine to keep it at its best operating temperature, just as the coolant in your car engine would.

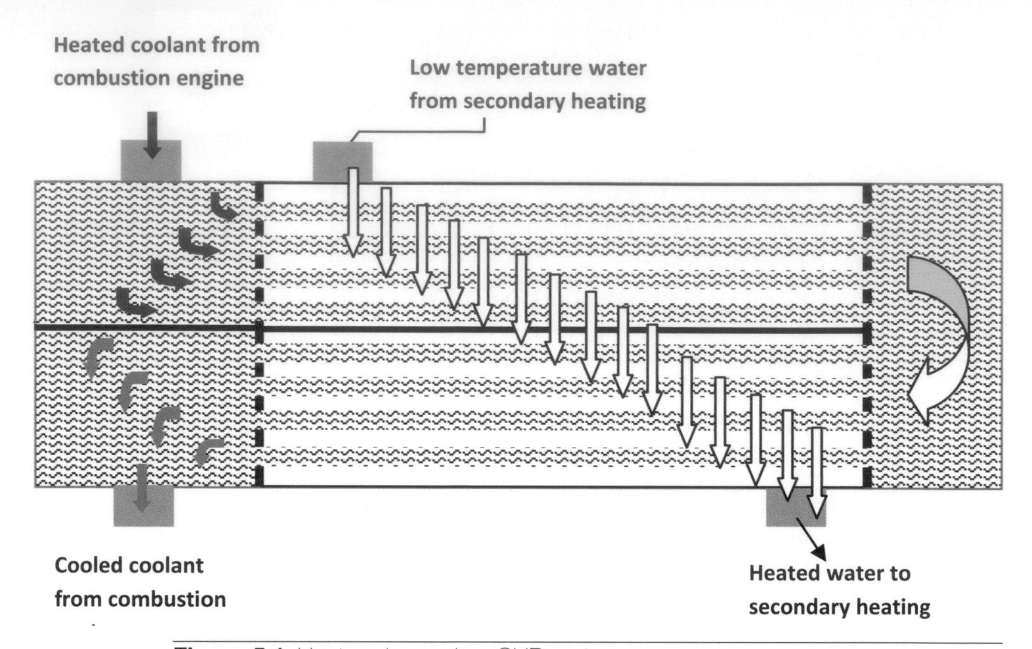

Figure 5.1 Heat exchange in a CHP system.

In a car the coolant is passed through a radiator which acts as a heat exchanger to cool it down, the coolant then passes back through the engine and is heated again as it keeps the engine cool. Of course the heat gained from the engine is simply lost into the atmosphere.

A CHP makes use of the heat from the coolant from the engine and instead of passing it through a radiator to cool it, a heat exchanger is used which transfers the heat from the coolant into the heating fluid that is used in a central heating or hot water storage system (Fig. 5.1).

This method makes very good use of the heat produced by the engine and at the same time it generates electricity. The problem is of course is that when no heat is required it is not very efficient to run the engine to produce electricity. In most cases the CHP is switched off and any electricity required is taken from the national grid. It is unlikely that a CHP would provide all of the energy required all of the time.

Another type of CHP uses a device called a Stirling engine to generate electricity using heat which has been produced by a heating boiler.

Stirling engine

One major benefit of a Stirling engine is that it is a completely sealed unit which has no emissions, the engine works using external extremes of heat to expand and contract a gas. This of course means that it can be used with any fuel.

The gas used inside the engine is normally helium, which expands when heated and contracts when cooled.

Stirling engines are very clever pieces of equipment which in many ways have been overlooked, probably due to the development of the internal combustion engine.

A simple explanation of how a Stirling engine works is as follows.

It consists of two cylinders each containing a piston. The cylinders are linked by a tube and one cylinder is heated and the other is cooled.

The tube which connects the two cylinders generally has a heat exchanger placed along it and this is often something as simple as a piece of mesh. The purpose of the heat exchanger is to help cool the helium gas as it is transferred from the hot cylinder to the cool one.

Although the engine has two cylinders it does not operate as you may expect with one piston being at the top of its stroke when the other is at the bottom, it is a little more complex than that.

As you can see from Figure 5.2 the pistons are linked to a fly wheel. When the hot piston is at the bottom of its stroke the cold piston is about midway, as the flywheel rotates clockwise and the hot piston begins its upward stroke and the cold piston is on its downward stroke. The gas is pushed from the hot cylinder through the heat exchanger into the cold cylinder. By the time the hot piston has reached the end of its upward stoke, the cold piston is halfway through its upward stroke pushing the gas through the tube via the heat exchanger again. When the cold piston has reached the end of

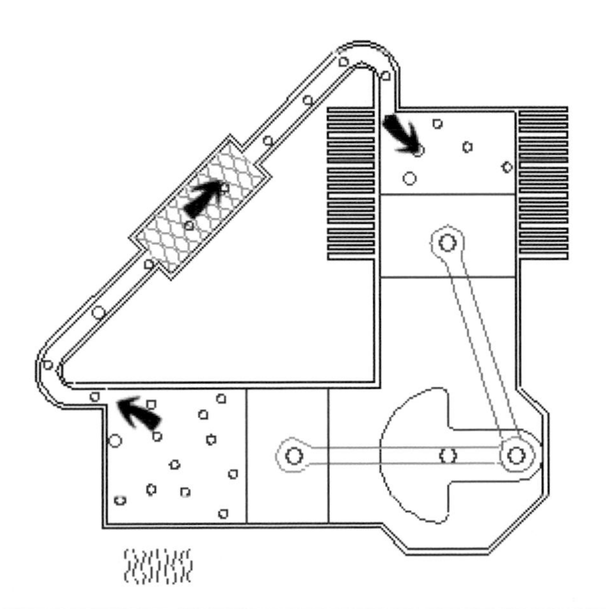

Figure 5.2 Cross section of typical Stirling engine.

its upward stroke the hot piston has begun its downward stoke and all of the gas has been pushed into the hot cylinder. The heat makes the gas expand very quickly and this of course forces the hot piston to move on its downward stroke, this is the part of the cycle which gives the engine its power. As the hot piston moves away from the top of its stroke so does the cold piston, this allows some of the expanded gas to pass through the heat exchanger again and into the cold cylinder. The cycle now begins again.

Figure 5.3 shows how the complete cycle works.

In many cases the pistons are magnetic and the cylinders contain coils of copper which generate electricity when the magnetic piston moves backwards and forwards through the copper coils.

New technology has been used which allows a Stirling engine to be fitted inside a boiler and it really does not matter which fuel is being used by the boiler to heat the water.

As the boiler is heating the water, the heat is also used to power a Stirling engine which in turn will generate electricity. In this type of boiler the electricity generated is really a by-product of the heating system, but the problem of course is that when there is no requirement for hot water there will be no electricity generated.

At present it is generally accepted that if the heat load required to make the system efficient is between 3 to 6kW for reasonably long periods, when the unit is fully operational it will be around 90 to 95% efficient.

I am sure that modern technology will improve on these figures as this new technology evolves. Unless this type of system is fuelled by a renewable energy such as a type of biofuel it could not really be classed as a renewable energy system as at the moment most of these systems are powered by gas or oil. There are still environmental benefits to be had whatever the type of fuel source as it is thought that this type of system saves around 60% on carbon emissions when compared to a conventional heating system.

Although this book is intended to explain different types of renewable energy I think that it is worth while adding a few more pieces of information about Stirling engines. I have provided some simple drawings to explain how the engine works, however they can be produced in many different ways and can even work using the heat from a cup of tea. The internet is a good source of information, www.howstuffworks.com or *the Stirling energy society* web site is worth a visit.

As already stated, the engine produces no emissions although the fuel used to heat the engine may. This will not generally add to the carbon emissions, particularly when used in a CHP, as the heat to operate the engine is also being used to heat water.

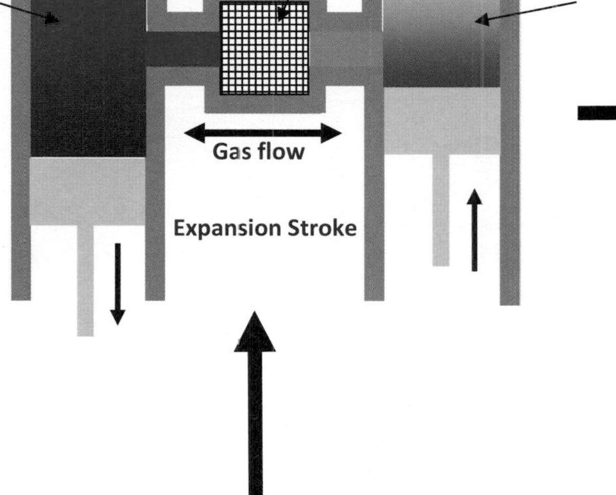

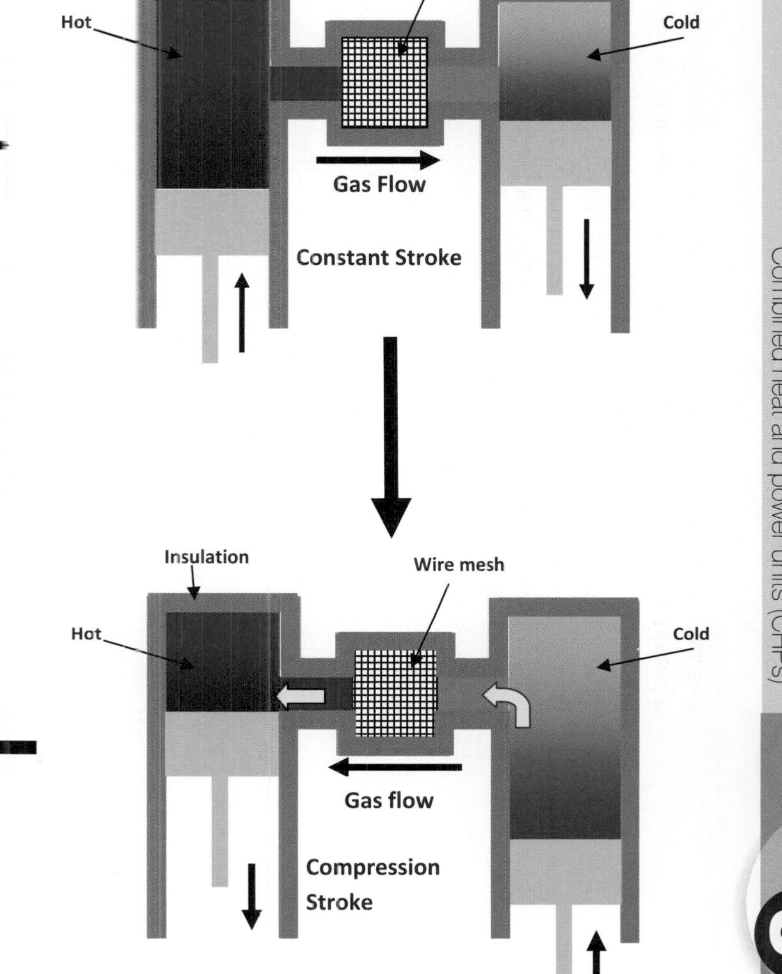

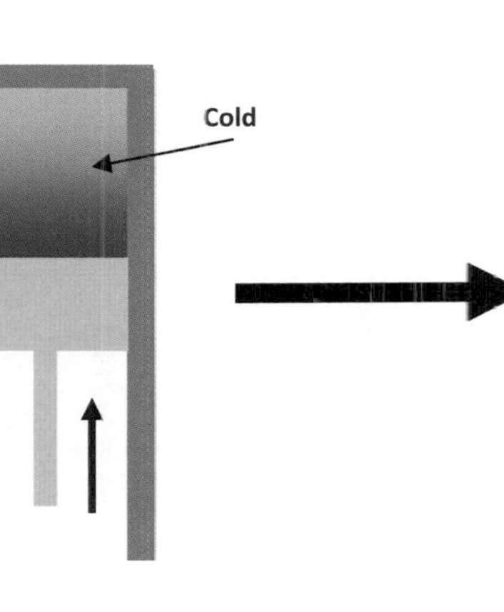

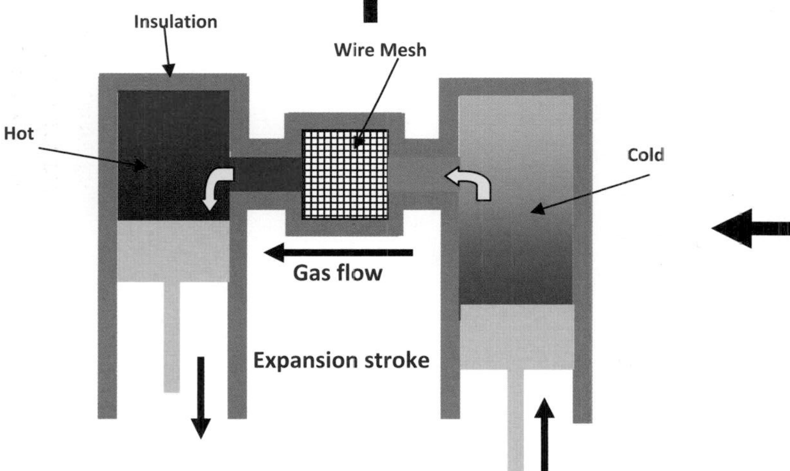

Figure 5.3 Stirling cycle.

Internal combustion engines use explosions within a cylinder to drive a piston, but this is a noisy process. A Stirling engine simply operates on the expansion of a gas and is relatively quiet, making it an ideal machine for use indoors. Stirling engines are also fully reversible, which makes them very useful for use in heat pumps to transfer heat from water to air or air to air. This will require the Stirling engine to be driven by a separate mechanical device. In this mode the Stirling engine will become a generator rather than a motor as it generates heat using the residual heat which is taken from the ground or the air. If the engine is reversed it will provide air conditioning rather than heating.

The Stirling engine does have one major disadvantage in that it does not produce as much power as an internal combustion engine. For this reason it is unlikely to be used in motor vehicles, but who knows what new technology will bring. Figure 5.4 is a typical micro combined heat and power unit.

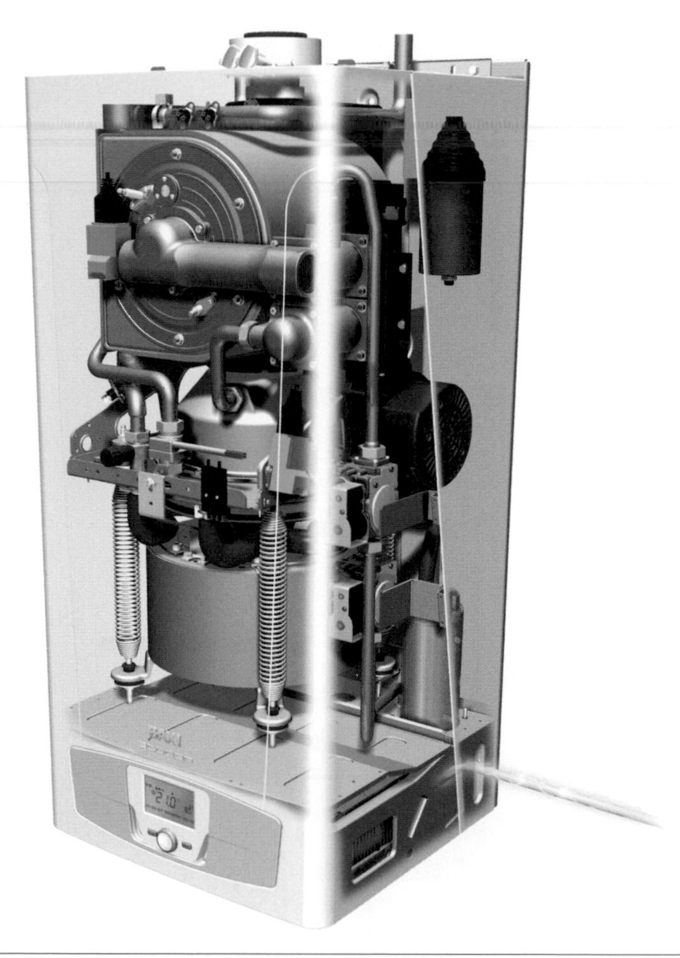

Figure 5.4 Combined heat and power unit.
Source: BDR Thermea

CHP units can also make use of fuel cells and although the first fuel cell was built in 1839 by Sir William Grove, there was little interest taken in it until the 1960s when the US government decided to use it as a safer option in their space programme. Fuel cells were used as a cheaper option to solar power and a safer option to nuclear power. Fuel cells are still used today to provide water and electricity in the space shuttle.

The problem arises with the cost. Fuel cells are very expensive to produce. However new technology and demand for this type of energy are likely to drive the cost down over a very short period of time. When these cells become less expensive micro CHP will probably be the best option for saving fuel in a domestic environment.

Index

T

U

V